Integrated Water Management

Integrated Water Management:

International Experiences and Perspectives

Edited by Bruce Mitchell

Belhaven Press
A Division of Pinter Publishers
London and New York

© The Editor and contributors, 1990

First published in Great Britain in 1990 by
Belhaven Press (a division of Pinter Publishers),
25 Floral Street, London WC2E 9DS

British Library Cataloguing in Publication Data
A CIP catalogue record for this book is available from the
British Library

ISBN 1 85293 026 8

Library of Congress Cataloging-in-Publication Data
Integrated water management : international experiences and
    perspectives / edited by Bruce Mitchell.
        p.   cm.
    ISBN 1-85293-026-8
    1. Water-supply—Management—International cooperation.
    I. Mitchell, Bruce, 1944–
    TD353.I526   1990
333.91′15—dc20   90-329
                                                        CIP

Typeset by Acorn Bookwork, Salisbury, Wiltshire
Printed and bound in Great Britain by Biddles Ltd, of Guildford and Kings Lynn

БСТ

Table of contents

$39.00

List of figures   vii
List of tables   ix
List of contributors   xi
Preface   xiii

1. Integrated water management   1
   Bruce Mitchell
2. Integrated water management in the United States   22
   Keith W. Muckleston
3. New Zealand water planning and management: evolution or   45
   revolution
   Neil J. Ericksen
4. Integrated water management strategies in Canada   88
   Daniel A. Shrubsole
5. Integrated water management in England   119
   Alan S. Pitkethly
6. Institutional arrangements for Japanese water resources   148
   Yoshihiko Shirai
7. Water resources and management in Poland   172
   Zdzislaw Mikulski
8. Integrated water management: the Nigerian experience   188
   Ademola T. Salau
9. Patterns and implications   203
   Bruce Mitchell
   Index   219

List of figures

Figure 3.1 The New Zealand environment: physical relief (3.1a); mean annual rainfall (3.1b); pre-European vegetation about AD 1800 (3.1c); land use 1989 (3.1d); population distribution 1986 (3.1e); and main water use sectors 1987 (3.1f)

Figure 3.2 Water and soil administration in New Zealand in 1972. (Source: Anon., 1984, p. 20)

Figure 3.3 Methods of implementation of the Land and Water Management Plan (LAWMAP) for the Upper Waitemata Harbour Catchment (Auckland). (Source: Auckland Regional Authority, 1983b, p. 12)

Figure 3.4 Radical change in resource management in New Zealand, 1986. (Source: Blakeley, 1988, p. 6)

Figure 3.5 Regional administration in 1988 for: water and soil resources in 20 catchment cum water boards (3.5a); natural and physical resources in 22 regional and united councils (3.5b); and proposed combined functions of 3.5a and 3.5b in 14 regional and resource management councils based on major catchments (3.5c)

Figure 4.1 Potential water-deficient regions of Canada

Figure 4.2 Existing and proposed boundaries for the Conservation Authorities of Ontario (1987)

Figure 4.3 Selected water management activities conducted under the Canada Water Act

Figure 4.4 The Federal Environmental Assessment Review Process: theory and practice

Figure 5.1 A chronology of major events since 1945

Figure 5.2 The River Authorities, 1963–73

Figure 5.3 The principal elements of the National Water Plan 1973. (Source: Water Resources Board, 1973, *The Water Resources of England and Wales*, 2 volumes, HMSO)

Figure 5.4 The Regional Water Authorities

Figure 6.1 The location map of the major rivers in Japan

Figure 6.2 A model of water use facilities—topographic patterns relations in the drainage basin context

Figure 6.3   General location map of the Kitakami River basin

Figure 6.4   The integrated dam management system of the Kitakami River

Figure 6.5   General location map of the Yahagi River basin

Figure 6.6   The relationship of standard and dynamic approaches in the Yahagi River method (after Harashima, 1987)

Figure 7.1   Water balance elements in Poland (after Gutry-Korycka, 1985)

Figure 7.2   Water usage in Poland, 1960–2000

Figure 7.3   Map of Poland showing the *vojevoidship* (provinces) and the division of Water Management Directorates (after General conception . . . , 1987)

Figure 8.1   Major rivers and River Basin Development Authorities' areas

List of tables

Table 3.1 Land and water planning and management statutes and powers to act

Table 5.1 The river water quality surveys (Source: Kinnersley, 1988, *Troubled Water: Rivers, Politics and Pollution*, Hilary Shipman, London, p. 207)

Table 7.1 Annual water usage in Poland, 1960–87

Table 8.1 Financial allocation to water supply provision

Table 8.2 Land utilization in Nigeria for the years 1951, 1976 and 1986

Table 8.3 Examples of dams already built, under construction or proposed for the future

List of contributors

Neil J. Ericksen, Department of Geography, University of Waikato, Hamilton, New Zealand

Zdzislaw Mikulski, Faculty of Geography and Regional Studies, University of Warsaw, Warsaw, Poland

Bruce Mitchell, Department of Geography, University of Waterloo, Waterloo, Ontario, Canada

Keith W. Muckleston, Department of Geography, Oregon State University, Corvallis, Oregon, U.S.A.

Alan S. Pitkethly, Department of Geographical Sciences, Huddersfield Polytechnic, Huddersfield, West Yorkshire, UK

Ademola T. Salau, Faculty of Social Sciences, University of Port Harcourt, Port Harcourt, Nigeria

Yoshihiko Shirai, Graduate Centre, Hyogo University of Teacher Education, Hyogo, Japan

Daniel A. Shrubsole, Department of Geography, University of Western Ontario, London, Ontario, Canada

Preface

This book was conceived after studying the concept of integrated water management, and the way in which it has been interpreted and implemented in a variety of countries. This period of study began for me in the late 1970s, when I participated as a member of an evaluation team assessing the experience with comprehensive river basin planning and management in Canada. During the 1980s, I had the opportunity to observe and examine related experiences in a variety of developed and developing countries while a visiting professor at universities, conducting research, or serving as a consultant to various agencies.

It became apparent that people in many countries are struggling with how to address problems that cut across elements of the hydrological cycle, that transcend the boundaries among water, land and environment, and that interrelate water with broader policy questions associated with regional economic development and environmental management. At a conceptual level, there is often agreement that it is necessary to define and tackle problems characterized by their interrelationships and linkages. Indeed, in the literature these types of problems are frequently referred to as metaproblems, wicked problems, or messes. Our traditional methods have often allowed us to solve some problems but have left us with other more significant issues to deal with.

The idea, then, was to examine experiences in a variety of countries to see how the question of 'integration' was being approached. It was recognized that successful approaches in one country or region could not automatically be transferred to another country or region, but it did seem that there was potential value in discovering if there were common issues and problems, and if there might be solutions that were generic in nature. Also, it was thought that the people in the best position to explore the experiences in a given country were individuals living and working in those countries.

In identifying candidate countries, a number of considerations were taken into account. First, the intent was to have a mix of countries which would reflect different physical conditions, different types of ideologies and different approaches to integrated water management. Second, the goal was to select countries in which there was a tangible and sustained attempt to deal with the issue of integration. Third, it was intended that each selected country would offer some positive examples of dealing with integration, even if there was still scope for improvement. A fourth concern was whether potential authors were available who could be depended upon to produce quality chapters within a specified time-frame.

Many more countries were possible candidates than could be included. The mix of countries represented in this book does provide a good cross-section relative to the criteria which were used in selecting the countries to be included. It should be stated, however, that if every author who had agreed to contribute a paper had completed a chapter, there would have been more chapters here. Nevertheless, it was assumed from the outset that a few authors might well not complete their chapters, and this too was kept in mind when people were invited to be contributors. Every reader undoubtedly can think of at least one other country which should be represented in a book such as this one. Perhaps such thinking will lead eventually to another volume which will help to complete the picture on a global scale.

Several people were particularly helpful in the preparation of this volume. I would like to thank a colleague, Andrzej Kesik, who provided advice regarding Poland. I would also like to express my appreciation to two graduate students at the University of Waterloo, Thomas Carlisle and Nigel Watson, who as research assistants provided invaluable assistance by helping in the reading of various versions of chapters and in suggesting modifications. I would also like to thank Iain Stevenson of Belhaven Press for his continuing support and assistance. Iain and I have worked together for over a decade on different book projects, and his enthusiasm and realism have always been appreciated.

Above all, I would like to thank the authors who prepared chapters for the book. They are all people with busy schedules, and their willingness to take on yet another writing task was and is appreciated. It is hoped that the insight they provide here about conditions, progress, successes and frustrations regarding integrated water management in their own countries will be of interest and help to readers.

Bruce Mitchell
Waterloo, Ontario
October 1989

CHAPTER 1

Integrated water management

Bruce Mitchell

1. Introduction

Integrated water management may be contemplated in at least three ways. First, it can imply the systematic consideration of the various dimensions of water: surface and groundwater, quantity and quality. The key aspect here is acceptance that water comprises an ecological system which is formed by a number of interdependent components. Each component (quantity and quality, surface and groundwater) may influence other components and therefore needs to be managed with regard to its interrelationships. At this level of integration, attention for management is directed to joint consideration of such aspects as water supply, waste treatment and disposal and water quality.

Second, integrated water management can imply that, while water is a system, it is also a component which interacts with other systems. In that respect, it directs us to address the interactions between water, land and the environment, recognizing that changes in any one may have consequences for the others. This perspective is broader than the first, and requires attention to, and expertise in, both terrestial and aquatic issues. At this second level, management interest becomes focused upon issues such as floodplain management, erosion control, nonpoint sources of pollution, preservation of wetlands and fish habitat, agricultural drainage and the recreational use of water.

A third and even broader interpretation is to approach integrated water management with reference to the interrelationships between water and social and economic development. At this level, the approach is on the scale recommended by the Brundtland Commission, with its stress upon the relationship between environment and economy. Here, the concern is to determine the extent to which water is both an opportunity for and a barrier against, economic development, and to ascertain how to ensure that water is managed and used so that development may be sustained over the long term. At this third level, interest turns to the role of water in producing hydroelectricity, in facilitating transportation of goods and in serving as an input to manufacturing or industrial production.

In addition to recognizing these three different possible interpretations of the meaning of 'integrated water management', it is important to recognize that the concept may be applied to different levels of analysis: normative, strategic and operational. At the normative level, attention is directed to decisions as to *what ought to be done*. At a strategic level, interest shifts to *what can be done*, whereas at the operational level the concern is with *what will be done*. Integration can and should occur at all of these levels, but it is important to be aware that in the management of water and related resources we often move from one level to the next.

In this book, the main focus is upon the integration of water with land and related environmental resources, with primary attention directed to the strategic level. At the same time, attention also is given to questions at normative and operational levels.

2. An approach with many applications

If integrated water management is interpreted to mean the idea of co-ordinating water, land and environmental resources in the context of catchments, it is an idea that has received considerable attention (Marshall, 1957; White, 1957; Weber, 1964; Ingram, 1973; McMillan, 1976; Schramm, 1980; Saha and Barrow, 1981; Wengert, 1981; Boisvert *et al.*, 1982; Falkenmark, 1985; Lundquist, 1985). Perhaps the best known model of integrating water and related resources is the Tennessee Valley Authority in the United States (Hodge, 1938; Ransmeier, 1942; Lilienthal, 1944; Clapp, 1955; Kyle, 1958; Hubbard, 1961; Selznick, 1966; Owen, 1973; Callahan, 1980). Established in 1933, the Tennessee Valley Authority was created to provide electricity from hydroelectric generation, to control flooding and to provide for transportation of goods. In addition, the TVA became involved in such diverse activities as rural planning, provision of housing, health care, libraries and recreation. Despite being referred to as an exemplary form of integrated resource management and development, the TVA has not been replicated in the United States or any other country (Clark, 1946).

Less well known but equally innovative in the United States are the watershed Conservancy Districts established in Ohio beginning in 1913. Included here are the Miami Conservancy District (Giertz, 1974) and the Muskingum Watershed Conservancy District (Browning, 1949; Craine, 1957; Jenkins, 1976). Other forms of integration of water and land in the United States have been soil conservation districts (Parks, 1952), special planning commissions (Clinkenbeard, 1973), regional authorities (Martin *et al.*, 1960; Hendrickson, 1981) and state-level initiatives (Blake, 1980; Allee, Dworsky and North, 1982).

Integrated water management has also occurred in other countries. In England and Wales, there has been an evolution towards a smaller number of river basin authorities with an increasing number of management functions (Funnel and Hey, 1974; Okun, 1977; Sewell and Barr, 1978; Kinnersley, 1988). Basin agencies were established in France during 1964 (Lamour, 1961; Harrison and Sewell, 1976). In Canada, the Ontario Conservation Authorities were created in 1946 (Pleva, 1961; Richardson, 1974), and other initiatives followed in other regions of the country (Dufour and Lemieux, 1978; Mitchell, 1983; Mitchell and Gardner, 1983). In New Zealand, attention has been directed to the interrelationship between land and water since before the Second World War (Howard, 1988), while in neighbouring Australia the concept of total or integrated catchment management received much attention beginning in the mid 1980s (Sewell, Handmer and Smith, 1985; Burton, 1986; Cunningham, 1986; Australian Water Resources Council, 1988; Mitchell and Pigram, 1989). In the developing countries, strong interest has also been shown in ways of considering the links between water and land. Examples are found in Nigeria (Faniran, 1972; Faniran, Akintola and Okechukwu, 1977; Adams, 1985), India (Bose, 1948; Kirk, 1950; Johnson, 1979; Saha, 1979) and the Philippines (Rondinelli, 1981).

Despite the general acceptance of the concept of integrating water and land management, progress in actual implementation has been hesitant and unsystematic. In part, this is a result of the presence of real obstacles to integration. It also reflects a situation in which participants are learning as they proceed, with no obviously correct model to follow. As a result, individuals are usually cautious and follow an incremental strategy in which they move forward slowly.

Numerous problems are encountered in implementing an integrated approach. Questions arise regarding what information, either existing or new, is needed to assist in planning and management decisions. There are other queries concerning the role of individual landowners and communities in preparing an integrated strategy. More generally, there are questions as to how to incorporate the general public into the process. For, without the interest and support of the local community, it is probable that efforts to implement an integrated approach will encounter difficulty. Questions also arise concerning the financial resources to support any initiatives. Will additional funds be provided to support an extended range of activities, and, if so, by whom? Or, conversely, is it assumed that gains in efficiency from an integrated approach will allow this activity to be introduced with existing or even reduced levels of funding?

The above points are all significant, and deserve careful attention. However, other more fundamental issues also need to be addressed. Specifically, what is meant by an integrated approach? Unless the concept is defined clearly, it will be difficult to establish goals and targets, and to monitor progress. Second, what are the necessary conditions to facilitate the introduction of an integrated approach? It is these two matters which are addressed in this opening chapter.

3. The scope of integrated water management

There is intuitive appeal for the idea of examining a broad array of variables and their interrelationships, in the context of water and land management. This appeal reflects recognition that many land-based activities have implications for water flows and quality. Furthermore, water can have significant implications for the land, through its capacity to trigger erosion, contribute to salinity and support wildlife. Thus, consideration of land and water through an integrated approach offers the possibility of addressing the dynamics of an ecological system, thereby ensuring that critical relationships are identified and managed.

Care must be taken regarding the scope of integrated water management. At a strategic level, it is appropriate and desirable to think *comprehensively*, where that is taken to mean trying to identify and consider the broadest possible range of variables which may be significant for coordinated management of water and associated land and environmental resources. However, at an operational level, maintaining a comprehensive perspective may create difficulties. Experience in many countries with the comprehensive approach at an operational level has indicated that it results in inordinately lengthy periods of time for planning and in plans which are often not sufficiently focused to be helpful to line managers.

The complexity and difficulty of trying to sustain a comprehensive approach at the operational level can often be overwhelming. Friend, Power and Yewlett (1974, xxiii) have noted that the more comprehensive the view taken in planning, the more planners find themselves dependent upon outcomes of other organizations, both public and private. Furthermore, they become more aware of all the subtle relationships—economic, social, political, ecological—which extend into other aspects of their activities and for which there is no well defined organizational structure.

As a partial solution, it is appropriate to recognize two levels in managing land and water resources. At the strategic level, a comprehensive approach should be used to ensure that the widest possible perspective is maintained. In contrast, at the operational level, a more focused approach is needed. To make the distinction between these two levels, it is suggested that *comprehensive* be used to describe the approach at the strategic level and that *integrated* be used to describe the approach at the operational level. The key distinction is that while an integrated approach maintains the interest in considering a mix of variables and their interrelationships, the focus is more selective or narrow than in the comprehensive approach. At the operational level, attention is directed toward a smaller number of variables which are believed to account for a substantial portion of the management problems.

In following the procedure outlined above, it should be possible to obtain the benefits of a comprehensive approach without becoming so entangled with a complex web of interrelationships that the management exercise literally disappears into a 'black hole', never to re-emerge. In that way, the time required for analysis and planning should be kept to a more reasonable length, and the tighter focus should result in a process which addresses the real needs of managers and users.

Another aspect deserves attention when adopting an integrated approach. Two options exist for developing a management plan. The first involves the identification of basic goals or directions as well as the activities needed to achieve those goals or directions. Here, the various individual participants and agencies determine how they might contribute to the achievement of common goals and directions. In such a process, it will also become clear where interests and values diverge, and therefore where negotiation will be necessary to address real and legitimately different interests and aspirations.

The second option involves the various individual participants and agencies identifying their own goals and directions. They then meet to see how, if at all, their diverse goals and directions can be interrelated and co-ordinated. This approach can create a major problem, as it focuses effort upon trying to reconcile individual participants' aspirations rather than concentrating upon broader considerations which reflect general social welfare. Individual and agency mandates and goals obviously always do exist, and it is not possible to ignore or forget about these. However, to let them establish the context for the integrated approach is similar to placing the cart before the horse. Practical experience suggests that a key to effective implementation is to identify common goals and objectives as well as the general direction judged to be desirable, and then to explore how individual organizations can contribute to them.

Several key points should be emphasized. First, if the planning and management exercise is to be productive, the scope of integrated water management must be well defined. Too often, people have had only a vague idea as to what 'integration' means, and/or there have been differing perceptions as to what it means. In such situations, it is not surprising that confusion has been frequent, and that many planning documents have been characterized by vagueness or fuzziness. Second, a comprehensive perspective is valuable for the initial review of a problem but should be followed at an operational level by an integrated approach which is more selective and focused. Third, common goals and activities need to be identified, so that participants can consider how they may contribute to them. These three characteristics by themselves will not resolve the problems of making the integrated approach effective, but they should remove or reduce many of the difficulties which have been encountered in the past.

4. The reality of boundary problems

Integration of water and land resources is usually supported for several reasons. First, as already mentioned, it is often argued that water problems have their origins on the land or from other economic and social activities. Second, public management agencies usually have fragmented and shared responsibilities either from one level of government to another (local, state, federal) or among agencies at the same level of government (water, agriculture, forestry, wildlife, minerals). As a result of these considerations, it is often accepted that integration will lead to co-operation and co-ordination which in turn will lead to overall improved effectiveness.

However, fragmented and shared responsibilities are realities which are always likely to exist. They become a barrier to, and a rationale for, integration. Eddison (1985, p. 149) has commented that the major management problems are always at the boundaries—or those points situated between states, between levels of government, among agencies, or among divisions within departments. He argued that the definitive resolution of boundary problems is not possible because at any time there are always several different and plausible approaches. Furthermore, from one viewpoint or another, every solution will have a weakness. He concluded that we need processes of administration to address the difficulties caused by boundary problems.

If boundary problems are significant and have to be overcome, it is essential to examine those aspects which are necessary elements in designing a strategy to achieve integration. Reorganizing the structure of public agencies is too simplistic an approach. Kellow (1985, p. 114) has maintained that we should judge the effectiveness of administration not by how many agencies are involved but rather by the effectiveness of the co-ordinating mechanisms. In Kellow's (1985, p. 118) opinion, 'rather than stress the degree of fragmentation which exists between the large list of agencies' it is more productive to concentrate upon 'an assessment of the adequacy of the co-ordinating mechanisms which exist' because that is 'the crucial test of institutional adequacy'.

5. Institutional arrangements

The term 'institutional arrangements' represents an area which is increasingly receiving attention from individuals in a variety of disciplines and professions (Gormley, 1987). The concept has been defined in different ways (Craine, 1971; Mitchell, 1975; Ingram et al., 1984; Fernie and Pitkethly, 1985). In this chapter, institutional arrangements are defined (after Mitchell, 1989, p. 245) as the combination of (1) legislation and regulations, (2) policies and guidelines, (3) administrative structures, (4) economic and financial arrangements, (5) political structures and processes, (6) historical and traditional customs and values and (7) key participants or actors.

Research on institutional arrangements has been characterized as being in its infancy, with description dominating analysis (Mitchell, 1989, p. 261).

Indeed, Ingram *et al.* (1984, p. 323) have characterized the research on institutional arrangements in water management as usually being un-illuminating as it is often little more than an annotated listing of public agencies, statutes, regulations, compacts and decisions. In such work, an image is created of a lifeless maze of check stations, passageways and barriers. In their view, 'form prevails over substance in such presentations, and little understanding of the dynamics of institutional operations or change is presented'.

To overcome these deficiencies, Ingram *et al.* (1984) urge that analysts should assemble information about the participants and their stakes, the resources to which they have access in pursuing their interests (legal rules and arrangements, economic power, prevailing values and public opinion, technical expertise and control of information, control of organizational and administrative mechanisms, political resources), and the biases in-herent in alternative decision-making structures.

In Section 4 the ubiquitous nature of boundary problems was stressed. Recognition of these boundary problems has led to interest in what Paterson (1986, p. 96) has referred to as the 'cult of co-ordination'. In this section it has been argued that there are a number of dimensions to institutional arrangements, each of which provides a stimulus or potential leverage point to foster integration and therefore, hopefully, to improve co-ordination. These leverage points must be viewed as interrelated, since no one by itself is capable of making integration occur. Indeed, reliance upon individual measures is probably the reason for Eddison's (1985, p. 149) conclusion that 'the familiar co-ordination devices do not really penetrate the problem'. In the following section, a framework is presented which identifies these key leverage points.

6. An analytical framework

The framework presented here can be used in two ways. In a *descriptive mode*, it should help to concentrate attention upon key events, decisions and people in a resource management situation. In a *prescriptive mode*, the guidelines which are included should provide direction for possible changes or modifications to a management situation. Recommendations for a specific situation will have to be custom-designed with regard to the characteristics pertinent to that situation.

The framework has six aspects: (1) context, (2) legitimation, (3) func-tions, (4) structures, (5) processes and mechanisms, (6) culture and attitudes. Each is presented in turn. It must be emphasized that there is not a linear relationship among them. In other words, for purposes of dis-cussion they are introduced in sequence, and there is logic for the order of presentation. However, it is stressed that each represents a necessary but not sufficient condition for the achievement of integration. Each represents a leverage point. In different situations, the relative importance of each aspect could vary. Thus, the framework is meant to be opportunistic in the sense that it identifies points at which leverage may be exerted to improve integration.

6.1. Context

In assessing the nature of, or opportunities of, integrating water and land resources it is essential to recognize the broader contextual aspects which alone or in combination may trigger a desire to pursue integration. As Ingram *et al.* (1984, p. 328) have observed, most decisions involving water resources are based on bargaining and compromise. The nature of the bargaining varies, however, depending on the diverse decision rules, traditions, legal authority and professional orientation appropriate to the situation.

For descriptive purposes, several considerations deserve attention. First, the *state of the natural environment* must be examined to determine if issues or problems associated with natural systems (pollution, drought, flooding, land degradation), are the stimuli for action. Second, prevailing *ideologies* need to be identified, as these will influence the choice of goals, objectives and strategies. Third, *economic conditions* often shape willingness to undertake new ventures. In difficult economic times, governments are frequently less inclined to undertake new initiatives. Fourth, existing *legal, administrative and financial arrangements* provide both opportunities and constraints. To understand the success or failure of integrative efforts, the analyst should have a solid understanding of these four elements. At this stage, no prescriptive guidelines are provided regarding these contextual aspects. Such guidelines are incorporated within subsequent subsections.

6.2. Legitimation

As boundary effects will inevitably be present, it is essential to identify: (1) the objectives of pertinent agencies, (2) the responsibility, power or authority of these agencies and (3) the rules for intervention and arbitration by higher level authorities when conflicts arise which cannot be resolved by the participants directly involved.

Knowing the objectives of each participating agency should help to identify both common and divergent interests. Clarity of objectives by itself will not ensure integration, since different agencies almost inevitably will have objectives which conflict. Furthermore, while clarity of objectives is desirable, if they are specified in great detail it is possible that an organizaion will be unreasonably constrained in being able to respond to changing conditions and in being able to negotiate or bargain with other participants. An argument can be made for some careful generalization in objectives so that flexibility is built in.

It is not sufficient to be aware of goals and objectives. In addition, guidance is required regarding the power or authority with respect to the goals, since often these may either be shared with or conflict with other organizations. When the relative power of different participants is not specified or made clear, the potential for stalemate is high, as various agencies jostle and manoeuvre for a lead position.

Even when objectives and powers are stated explicitly, there is still a strong probability of conflicts between participating agencies as a result of boundary effects. Thus, it is important to identify who has the right and duty to intervene to resolve such impasses. In democratic societies, this task is often allocated to the elected person who is responsible for a department or ministry, or in the last resort, to the leader of the governing party. In reality, however, such people are usually too preoccupied with broader issues and are not well informed about the details of land and water management. Thus, it becomes desirable to incorporate a mechanism into the governmental structure to facilitate resolution of legitimate differences of opinion.

Paterson (1986, p. 101) has outlined in vivid terms the problems which can arise when a group of participants with equal or undifferentiated responsibilities are brought together in an effort to integrate and co-ordinate their interests and activities. He argued that in normal situations the practical arrangements to deal with overlapping responsibilities are left to bureaucratic measures. When a group of officials meets to discuss co-ordination, each person comes with a mandate which addresses a specific purpose. Each person does not have the power to compromise the primary mandate of his or her agency, and no or few resources to trade. The solution is usually to seek a compromise between the conflicting primary objectives rather than by choosing between them. In Paterson's view, the outcome is that 'instead of a choice to do one thing or another, both will usually be done badly'.

Paterson described the above situation as a 'co-ordination trap' and argued that it is necessary to develop a mechanism which will stimulate one of the participants to escalate the issue or to elevate the problem to the attention of a central authority. He believed that the best means to accomplish this is 'to create a single statutory focus of responsibility for each major class of overlap or interaction, where these are explicitly identified'. He also recognized, however, that many interactions only arise in the course of events. For those situations, he suggested that 'the best that can be done is to vest the responsible authority with a concurrence power over any primary activity of another party to the discussion that imposes a negative externality on the first's' (Paterson, 1986, p. 101).

Statute or *legislation* is one means by which to identify the objectives, powers and conflict resolution mechanisms for resource management.

Paterson's argument for a single statutory focus for each major class of overlap or interaction would represent a significant and positive move towards addressing boundary problems.

Many countries already have statutory power to achieve a number of aspects of land and water integration. However, without *political commitment* to use the existing legislative authority, there is a real danger that even the most carefully and astutely crafted legislation will not be sufficient. Ideally, those concerned with achieving integration of land and water need to cultivate and sustain political commitment for this purpose. Without such commitment, the problems identified in Subsection 6.6 may be capable of blocking moves towards integration even when a single powerful statutory base has been put into place.

A third mechanism for legitimation can be *administrative or bureaucratic policy* developed within a department. It is normally within the discretionary power of senior public servants to direct their departmental activities towards co-ordination and integration of activities with other public agencies. Unfortunately, as a result of the nature of vertical and horizontal fragmentation, there is usually little incentive for senior public servants to think in those terms, despite Paterson's (1986, p. 98) observation that the working hours of many public officials are dedicated to co-ordination. Too often, administrative directives provide a veneer of support for integration, while the 'hidden' or 'unofficial' agenda supports the furtherance of agency interests.

Thus, the likelihood of effective integration is not great unless the concept has been given legitimacy. A statutory basis offers the most enduring support for integration, but the probability of real achievement increases substantially when a statutory base is combined with strong political commitment. Administrative or bureaucratic directives are not likely to provide a strong degree of legitimacy because too many other elements can undermine them (Subsection 6.6). Without explicit legitimation of some form, any initiative to encourage integration and co-ordination faces an arduous, uphill struggle.

6.3. Functions

Once the objectives, power and rules for conflict resolution have been established, the management functions need to be considered. These should be linked explicitly to the legitimation (Subsection 6.2) and to the structures (Subsection 6.4). A key concern should be to determine which management functions should be assigned to which scale or hierarchy (local, state, federal). Some functions lend themselves to the local level (e.g., water supply) whereas others are more appropriate at the state or national level (e.g., pollution regulation). A guiding principle is that the functions should be allocated whenever possible to the scale or hierarchy that has the closest relationship to those people receiving the service or product.

Management functions can be considered in at least two categories. *Generic functions* involve (1) data collection, (2) planning, (3) regulation, (4) development, (5) monitoring and (6) enforcement. *Substantive functions* are specific to a resource or sector. Substantive management functions of water management include (1) supply, (2) sewage treatment and (3) pollution control. The substantive functions concerning the linkages between land and water include (1) floodplain management, (2) erosion control, (3) drainage and (4) wetlands. At the level where the linkages among water, environment and economy are addressed, the substantive functions would include (1) navigation, (2) hydroelectricity production, (3) recreation and tourism and (4) industrial production. Comparable substantive management functions can be identified for other resources such as agriculture, forestry and wildlife. The purpose should be to align functions as effectively as possible to various scales or hierarchies.

Different mixes of scales, generic and substantive functions will be appropriate in different situations. A basic concern should be to examine the alignment of these features in a systematic manner, and not to assume that the current pattern which has evolved over time is necessarily either the most systematic or sensible for current conditions. It may well be that the alignment made excellent sense at the time decisions were taken, but in many instances those decisions were made decades previously when the context and conditions were significantly different.

6.4. Structures

Structures should normally have an explicit relationship with functions. Whereas 'appropriate structure does not guarantee good performance, . . . inappropriate structure is a virtual guarantee of sub-standard performance' (Hunter District Water Board, 1982, p. 24).

Paterson (1986, p. 98) has explained that a key structural element inhibits integration and inter-agency co-operation. In his view, in addition to the absence of markets for most public sector products and services, a major barrier is the necessity for operating the public sector as a single unit for purposes of public policy, but as a collection of substantially independent units for purposes of management.

In an earlier study, Paterson (1985, p. 141) addressed this dilemma with regard to the administration of land and water. He argued that everyone could agree that land and water management should be managed as a whole, but nobody has suggested how this can be done effectively. Having recognized the interdependence of land and water, however, he indicated that acceptance of that holistic character would lead ultimately to one big bureaucracy responsible for the world. In his view, such an arrangement is not politically acceptable, nor administratively feasible. The challenge, in his opinion, is to consider 'something which artificially gives us separability at some level lower than the global scale'. As soon as that position is taken, then at some point boundaries must be drawn which will break up the integrity of the holistic land-water system.

Paterson (1985, p. 142) adds another complication to the process of determining appropriate structures by reminding us that a key to motivation is to divide organizational units so that each one has a well defined function to perform as well as maximum autonomy to perform its task. Acceptance of this argument leads increasingly towards more fragmentation and division of responsibility.

Considerable discussion has occurred regarding the most appropriate organizational structures for integrated land and water management. Alternative forms are available. If the alternatives are viewed as points along a continuum, numerous possibilities are available. At one end of the continuum would be reliance upon fewer, larger, centralized, multiple-function organizations. The advantages of this alternative are usually viewed as economies of scale and realization of co-ordination through centralized control. At the other end of the continuum would be the use of a larger number of smaller, decentralized, special-purpose agencies. The advantages here are perceived to be organizations which are closely aligned to the constituency being provided with products and services. In addition, this approach incorporates a certain amount of tension as the different special-purpose agencies monitor and scrutinize each others' activities (Parker and Penning-Rowsell, 1980, p. 19).

In selecting an organizational structure, several considerations should be remembered. First, there never will be a perfect match between functions and structural form. Second, regardless of the structure chosen, boundary problems will emerge. The nature of the structure will simply determine whether the boundary problems occur among units within a larger agency or among a number of smaller, more specialized agencies. Third, the two alternatives described above represent the end points on a continuum. Various permutations and combinations could be constructed, and it is misguided to think of the alternatives in 'either/or' terms. Fourth, it is not necessarily appropriate to look for a single structure to handle all aspects of the management problem. As generic and substantive functions are allocated to different scales, it may turn out that different structures are appropriate for different scales.

Several other considerations deserve attention. Current social values place considerable weight upon *accountability* in government and policy making. In designing organizational structures, it is appropriate to consider how accountability might be built into the structures. This consideration is often considered with reference to the processes and mechanisms which are used in conjunction with the structures (Subsection 6.5). Interest is also often given to the concept of *flexibility*. In other words, in the design of organizational structures, consideration should be directed towards how an organization can be made capable of responding to changing or uncertain conditions. In some situations, this could lead to a conscious decision to have relatively impermanent structural forms so that they could be re-arranged as conditions warranted.

6.5. Processes and mechanisms

No matter how carefully considered and designed, the legitimation, functions and structures are unlikely to fit together perfectly. The presence of loose ends because of imperfect fits suggests the need for processes and mechanisms to facilitate bargaining, negotiating and mediating at the boundaries.

Processes may be used at the *political* and *bureaucratic* levels. At the political level, *interministerial councils* are often used to ensure that pertinent Cabinet Ministers meet to share views. *Select committees* may also be used when it is thought that a specific problem requires investigation by a group representing a cross-section of the political ideologies prevailing in a state or country.

At the bureaucratic level, numerous mechanisms have been used to facilitate integration and co-ordination. *Interdepartmental committees* are frequently used to provide a forum for exchanging information and ideas about ongoing and proposed activities. These structures are occasionally given the status of *commissions*. A common weakness of such committees is that they are given little power and therefore are placed in a weak position to effect change. Paterson (1986, p. 101) has aptly characterized them as the 'interdepartmental committee of fudges and compromises' which are ineffective not because of the inability of the participants but as a result of the fundamental constraints associated with their real authority.

Task forces are also used, and are usually established for a specific purpose and a set period of time. *Review procedures* in which government agencies circulate plans or proposals to related agencies for comment are also used. These four mechanisms are not mutually exclusive, as two or more can be used simultaneously.

Many *informal mechanisms* are used to realize integration. Professional counterparts in different agencies can telephone one another to share information or to solicit reaction to ideas. Informal discussions before or after interdepartmental committee meetings often have more substance and accomplish more than the extended and often ritualistic discussion which occurs around the interdepartmental committee table. Individuals will arrange to discuss matters of mutual concern over lunch or during work breaks. Thus, many mechanisms which do not appear on any organization charts or strategic plans exist and are used regularly. At the same time, many informal actions can be invoked to block or reduce integration. These will be considered in Subsection 6.6.

If integration is truly sought, then provision should be made to ensure that the viewpoints of the general community and of individuals are fed into planning and management exercises. In other words, some 'bottom up' elements are desirable as a counterpoint to much of the 'top down' orientation which characterizes resource management decision making. A 'bottom up' orientation leads to public participation in its various forms, and can range from consultative committees, advisory groups and public meetings, to surveys of the general public.

Whichever mix of mechanisms is chosen, a variety of processes can be drawn upon to pull together diverse viewpoints. *Regional planning* processes seek to incorporate a diversity of interests. *Benefit cost analysis* and *environmental impact assessment* also force attention towards an array of considerations. *Public participation* as a process serves to integrate different points of view. Each of these processes has relative merits. Consequently, as with the mechanisms, it is usually appropriate to determine which combination will best meet the needs in a given situation.

6.6. *Organizational culture and participant attitudes*

Ultimately, integration, co-operation and co-ordination depend to a significant extent upon the willingness of the participants to make them happen. As Kellow (1985, p. 121) has remarked, 'what is unwritten—the degree of co-operation and goodwill . . .—is more important than what is written down'. People who are inclined to co-operate and are enthusiastic can often make a poor system work well. Conversely, a well-designed system may falter if the participants are determined not to work with each other.

Unfortunately, there are usually few explicit incentives for integration and co-ordination. Vertical and horizontal fragmentation creates an environment in which rewards usually accrue to those who concentrate upon, indeed defend, their own areas of interest. As Eddison (1985, p. 150) remarked, 'much of the so-called conflict is nothing but self-indulgent skirmishing, deriving out of a system which has provided little or no alternative. Organisational skirmishing is a function of the culture of our organisations and nothing more'. It is this 'organizational culture' (Frost *et al.*, 1985) which encourages most public agencies to look to their own interests first and societal welfare second.

In addition to structural aspects which foster parochialism, there are the reactive effects of individuals. As O'Riordan (1976, p. 65) observed, decision making for resource management often has little to do with organization, statutory guidelines and co-ordinating arrangements. Rather, it has much more to do with the outcome of the determination, vision, indifference, antagonisms and bloody-mindedness of particular individuals who are in positions of influence. Thus, the combination of organizational culture, personalities and participants' attitudes can pose a major obstacle to integration and co-operation. That is reality.

Whereas society may gain through more co-ordination and co-operation among public agencies, some individuals will perceive themselves as becoming losers through reduction in or loss of authority, shrunken empires and reduced leverage or influence. In such situations, lip service may be given in support of integration, but in practice low-risk strategies such as delay, systematic disinformation and minor sabotage will be utilized to hinder its implementation.

Identifying the characteristics of the organizational culture and the participants' attitudes regarding disincentives and incentives for integration therefore becomes important. Through fuzzy legitimation, unclear functions and cumbersome structures, an organizational culture develops which creates real barriers to integrated and co-operative effort. In addition, professional biases and personal ambitions generate a hidden agenda which may thwart integration.

In a positive sense, a strong declaration about the need for integration on the political level will give a desirable signal to those responsible for implementing integration. A systematic linking of functions to different hierarchies or scales will help to pinpoint areas in which integration could be pursued. Careful design of organizational structures should ensure that diverse viewpoints are brought together if the design explicitly recognizes a role for landowners and resource users as well as the public servants in developing policy and programmes.

Concerning participants' attitudes, it is important to recall that '. . . most decisions involving water resources are made on the basis of bargaining, negotiation and compromise' (Ingram *et al.*, 1984, p. 328). Unfortunately, education programmes in resource management and in-career educational programmes appear to devote little time to developing the bargaining skills of resource managers. It seems to be assumed that bargaining skills will be gained through the experience of being involved in resource management. Such an approach is inefficient, as it requires the accumulation of many years of experience to develop good bargaining skills. An alternative approach would be to teach skills pertinent to bargaining, negotiation and compromise.

Numerous approaches and books have been prepared, but one of the better introductions is *Getting to Yes* by Fisher and Ury (1981). These authors outline a series of guidelines in a simple, straightforward and lively manner. For example, they argue that it is best to separate the characteristics of the people involved from the characteristics of the problem. They believe that too often people react to the people rather than concentrating upon the problems. They also suggest that interests rather than positions should be the focus of attention. They have concluded that different participants often turn out to have similar interests even though their public positions may appear to be divergent. These and other practical suggestions could help to improve bargaining skills, and could thereby shape participants' attitudes to be more positive towards integration if they believe that it is possible for everyone to 'get to yes' and enjoy some real gains.

Notwithstanding the optimism of Fisher and Ury, it is essential to appreciate that good bargaining skills will not be sufficient to achieve integration. Points will be reached beyond which individuals will not compromise and, when that happens, a stand-off or stalemate can occur. In such circumstances, it is important in the legitimation component to have clear guidelines as to who has the responsibility and power to intervene and break a deadlock.

As with processes and mechanisms (Subsection 6.5), being aware of the informal activities which can promote or hinder integration is important. In a positive sense, action can be taken to arrange informal meetings to discuss matters of mutual interest and concern, to cultivate a network of key contacts and sources of information, to place people on key committees and to befriend officials at various scales or hierarchy of government.

In a negative sense, deliberate delay of responses, withholding of information and other similar methods can be used to frustrate efforts of realizing integration and co-operation.

Ultimately, a key consideration is the recognition that organizational culture and participant attitudes can be crucial to the success of an integrated approach. Since many disincentives regarding integration exist, it is of fundamental importance that the 'human dimension' be given equal consideration relative to legitimation, functions, structures and processes/ mechanisms.

7. General implications

If co-ordinated management of water and land resources is to be achieved, the scope of a holistic approach must be carefully thought through. It is recommended here that a two-stage strategy be used. At a strategic level, a *comprehensive viewpoint* is desirable where that implies scanning the widest possible range of issues and variables. At an operational level, however, a more focused approach should be utilized. For purposes of differentiation, it is suggested that at the operational level this be called an *integrated approach*. The implication is that, at this level, attention concentrates upon those issues and variables judged to be the most significant. Thus, a *bounded holistic perspective* is achieved which should lead to findings and recommendations which are both timely and related explicitly to the needs of managers.

In developing this bounded holistic perspective, definition by the various participating individuals and organizations of a set of common goals and objectives is important. With those established, individuals and organizations can determine how they might contribute to the realization of these common goals and objectives. Without an overriding sense of common purpose or direction, it is unlikely that co-ordination will be achieved.

Boundary problems will always exist in natural resource management. No matter how much care is taken in allocating specific objectives and responsibilities, some overlap and conflict will emerge since society has multiple values, not all of which are compatible. Thus, the objective should be to devise institutional arrangements which will minimize the associated problems.

In developing institutional arrangements, we should recognize that numerous options are available, all with strengths and weaknesses. Since every option will be imperfect, attention should be directed towards identifying the mix of leverage points which can be used collectively to achieve co-ordination of water and land management. After recognizing the general *context* within which resource policies and decisions exist, it is suggested here that there are five key leverage points—*legitimation, functions, structures, processes and mechanisms, organizational culture and participant attitudes*. Each one of these is a necessary but insufficient condition for integration.

The five leverage points should not be viewed as a sequence of elements in which a manager mechanically moves from one to another in some predetermined order. It is more realistic to view each of these leverage points as a critical element in achieving integration, and to understand that, in different circumstances, their relative importance can vary. In that sense, the framework identifies opportunity targets for the resource manager. Since there are usually multiple causes for management problems, believing that there should be multiple responses is reasonable. The framework emphasizes the need to consider simultaneously a mix of responses. Furthermore, the framework stresses the significance of the human dimension, and reminds us that ultimately people will be responsible for implementing any policy or programme. As a result, any design of institutional arrangements should incorporate some 'bottom up' features to counterbalance 'top down' directives and initiatives.

In the subsequent chapters of this book, there is an opportunity to see how the concept of integrated water management has been conceived and implemented in a number of countries. In the final chapter, the general lessons and implications from this diverse experience will be considered.

Notes

This chapter is a modified version of the following publication: B. Mitchell, 1987, *A Comprehensive-Integrated Approach for Water and Land Management*, Occasional

Paper No. 1, Centre for Water Policy Research, University of New England, Armidale, NSW, Australia.

References

Adams, W. M., 1985, 'River basin planning in Nigeria', *Applied Geography*, **5**, pp. 297–308.

Allee, D. J., Dworsky, L. B. and North, R. M. (eds.), 1982, *Unified River Basin Management—Stage II*, American Water Resources Association, Minneapolis.

Australian Water Resources Council, Australian Environment Council, and Australian Soil Conservation Council, 1988, *Proceedings of the National Workshop on Integrated Catchment Management*, Australian Water Resources Council Conference Series no. 16, Victoria Department of Water Resources, Melbourne.

Blake, N. M., 1980, *Land into Water—Water into Land: A History of Water Management in Florida*, University Presses of Florida, Tallahassee.

Boisvert, A., Levert, F., Lussier, M.-J. and Sénécal, P., 1982, 'Pour une gestion globale de l'eau', *Eau du Québec*, **15** (3), pp. 273–7.

Bose, S. C., 1948, *The Damodar Valley Project*, Phoenix, Calcutta.

Browning, B., 1949, 'The Muskingum story', *Journal of Soil and Water Conservation*, **4** (1), pp. 13–16, 20.

Burton, J. R., 1986, 'The total catchment concept and its application to New South Wales', *Hydrology and Water Resources Symposium 1986: River Basin Management*, Institution of Engineers, Australia, Barton, ACT, pp. 307–11.

Callahan, N., 1980, *TVA: Bridge Over Troubled Waters*, A. S. Barnes and Co., South Brunswick, NJ and New York.

Clapp, G. R., 1955, *The TVA: An Approach to the Development of a Region*, University of Chicago Press, Chicago.

Clark, W. C., 1946, 'Proposed "Valley Authority" legislation', *American Political Science Review*, **40** (1), pp. 62–70.

Clinkenbeard, H. E., 1973, 'Integrated natural resources into area wide and local planning: the southeastern Wisconsin experience', *Journal of Soil and Water Conservation*, **28** (1), pp. 32–5.

Craine, L. E., 1957, 'The Muskingum Watershed Conservancy District: a study of local control', *Law and Contemporary Problems*, **22**, pp. 378–404.

Craine, L. E., 1971, 'Institutions for managing lakes and bays', *Natural Resources Journal*, **11**, pp. 519–46.

Cunningham, G. M., 1986, 'Total catchment management—resource management for the future', *Journal of Soil Conservation*, New South Wales, **42** (1), pp. 4–5.

Dufour, J. and Lemieux, G. H., 1978, 'L'aménagement des berges, ravins et monts urbains dans la conurbation du Haut-Saquenay: vers un espace plus fonctionnel', *Cahiers du Géographie de Québec*, **22** (57), pp. 421–35.

Eddison, T., 1985, 'Managing an ecological system 5: reforming bureaucracy', *Australian Quarterly*, **57** (1&2), pp. 148–53.

Falkenmark, M., 1985, 'Integration in the river-basin context', *Ambio*, **14**, 118–20.

Faniran, A., 1972, 'River basins as planning units' in K. M. Barbour (ed.), *Planning for Nigeria: A Geographical Approach*, Ibadan University Press, Ibadan, pp. 128–54.

Faniran, A., Akintola, F. W. and Okechukwu, G. C., 1977, 'Water resources development process and design: case study of the Oshun river catchment' in A. L. Mabogunje and A. Faniran (eds.), *Regional Planning and National Development in Tropical Development*, Ibadan University Press, Ibadan.

Fernie, J. and Pitkethly, A. S., 1985, *Resources: Environment and Policy*, Harper and Row, London.

Fisher, R. and Ury, W., 1981, *Getting to Yes*, Business Books, London.

Friend, J. K., Power, J. M. and Yewlett, C. Y. L., 1974, *Public Planning: the Inter-Corporate Dimension*, Tavistock Publications, London.

Frost, P. J., Moore, L. F., Louis, M. R., Lundberg, C. C. and Martin, J. (eds.), 1985, *Organizational Culture*, Sage Publications, Beverly Hills.

Funnell, B. M. and Hey, R. D. (eds.), 1974, *The Management of Water Resources in England and Wales*, Saxon House, Farnborough.

Giertz, J. F., 1974, 'An experiment in social choice: the Miami Conservancy District, 1913–1922', *Public Choice*, **19** (Fall), pp. 63–76.

Gormley, W. T., 1987, 'Institutional policy analysis: a critical review', *Journal of Policy Analysis and Management*, **6**, pp. 153–69.

Harrison, P. and Sewell, W. R. D., 1976, 'La ré-organization économique et régionale de la gestion des eaux en France', *Cahiers de Géographie de Québec*, **20** (49), pp. 127–42.

Hendrickson, K. E., 1981, *The Waters of the Brazos: A History of the Brazos River Authority, 1929–1979*, The Texian Press, Waco, Texas.

Hodge, C. L., 1938, *The Tennessee Valley Authority: A National Experiment in Regionalism*, American University Press, Washington, DC.

Howard, R., 1988, 'A New Zealand perspective on integrated catchment management', *Working Papers for the National Workshop on Integrated Catchment Management*, Australian Water Resources Council, Melbourne, pp. 101–31.

Hubbard, P. J., 1961, *Origins of the TVA: the Muscle Shoals Controversy, 1920–1932*, Vanderbuilt University Press, Nashville, Tennessee.

Hunter District Water Board, 1982, *Annual Report, 1981–1982*, Hunter District Water Board, Newcastle, New South Wales, Australia.

Ingram, H., 1973, 'The political economy of regional water institutions', *American Journal of Agricultural Economics*, **55** (1), pp. 10–18.

Ingram, H. M., Mann, D. E., Weatherford, G. D. and Cortner, H. J., 1984, 'Guidelines for improved institutional analysis in water resources planning', *Water Resources Research*, **20**, pp. 323–34.

Jenkins, H. A., 1976, *A Valley Renewed: The History of the Muskingum Watershed Conservancy District*, Kent State University Press, Kent, Ohio.

Johnson, S. H. *et al.*, 1979, 'Improving water management in the Indus basin', *Water Resources Bulletin*, **15** (2), pp. 473–95.

Kellow, A. J., 1985, 'Managing an ecological system 2: the politics of administration', *Australian Quarterly*, **57** (1&2), pp. 107–27.

Kinnersley, D., 1988, *Troubled Water: Rivers, Politics and Pollution*, Hilary Shipman, London.

Kirk, W., 1950, 'The Damodar Valley—Valles Optima', *Geographical Review*, **40**, pp. 415–43.

Kyle, L. H., 1958, *The Building of TVA*, Louisiana State University Press, Baton Rouge, Louisiana.

Lamour, P., 1961, 'Land and water development in Southern France' in H. Jarrett (ed.), *Comparisons in Resource Management*, Johns Hopkins Press, Baltimore, pp. 227–50.

Lilienthal, D. E., 1944, *T.V.A.: Democracy on the March*, Harper and Bros., New York.

Lundquist, J., Lohm, U. and Falkenmark, M. (eds.), 1985, *Strategies for River Basin Management (Environmental Integration of Land and Water in a River Basin)*, D. Reidel Publishing Co., Dordrecht, The Netherlands.

Marshall, H., 1957, 'The evaluation of river basin development', *Law and Contemporary Problems*, **22**, pp. 237–57.

Martin, R. C., Birkhead, G. S., Burkhead, J. and Munger, F. J., 1960, *River Basin Administration and the Delaware*, Syracuse University Press, Syracuse, New York.

McMillan, M., 1976, 'Criteria for jurisdictional design: issues in defining the scope and structure of River Basin Authorities and other decision-making bodies', *Journal of Environmental Economics and Management*, **3** (1), pp. 46–68.

Mitchell, B. (ed.), 1975, *Institutional Arrangements for Water Management: Canadian Experiences*, Publication Series no. 5, Department of Geography, University of Waterloo, Waterloo, Ontario.

Mitchell, B., 1983, 'Comprehensive river basin planning in Canada: problems and opportunities', *Water International*, **8** (4), pp. 146–53.

Mitchell, B., 1989, *Geography and Resource Analysis*, Longman, London, second edn.

Mitchell, B. and Gardner, J. S. (eds.), 1983, *River Basin Management: Canadian Experiences*, Publication Series no. 20, Department of Geography, University of Waterloo, Waterloo, Ontario.

Mitchell, B. and Pigram, J. J., 1989, 'Integrated resource management and the Hunter Valley Conservation Trust, N.S.W., Australia', *Applied Geography*, **9**, pp. 196–211.

Okun, D. A., 1977, *Regionalization of Water Management: A Revolution in England and Wales*, Applied Science Publishers, London.

O'Riordan, T., 1976, 'Workshop on resource management decision making', *Area*, **8**, p. 65.

Owen, M., 1973, *The Tennessee Valley Authority*, Praeger, New York.

Parker, D. J. and Penning Rowsell, E. C., 1980, *Water Planning in Britain*, George Allen and Unwin, London.

Parks, W. R., 1952, *Soil Conservation Districts In Action*, Iowa State College Press, Ames, Iowa.

Paterson, J., 1985, 'Managing an ecological system 4: reforming the tariff', *Australian Quarterly*, **57** (1&2), 139–47.

Paterson, J., 1986, 'Co-ordination in government: decomposition and bounded rationality as a framework for "user friendly" statute law', *Australian Journal of Public Administration*, **45**, pp. 95–111.

Pleva, E. G., 1961, 'Multiple purpose land and water districts in Ontario', in H. Jarrett (ed.), *Comparisons in Resource Management*, Johns Hopkins Press, Baltimore, pp. 189–207.

Ransmeier, J. S., 1942, *The Tennessee Valley Authority: A Case Study in the Economics of Multiple Purpose Stream Planning*, Vanderbuilt University Press, Nashville, Tennessee.

Richardson, A. H., 1974, *Conservation by the People*, University of Toronto Press, Toronto.

Rondinelli, D. A., 1981, 'Applied policy analysis for integrated river basin development programs: a Philippines case study' in S. K. Saha and C. J. Barrow (eds.), *River Basin Planning: Theory and Practice*, John Wiley, Chichester, pp. 285–323.

Saha, S. K., 1979, 'River basin planning in the Damodar valley of India', *Geographical Review*, **69** (3), pp. 273–87.

Saha, S. K. and Barrow, C. J. (eds.), 1981, *River Basin Planning: Theory and Practice*, John Wiley, Chichester.

Schramm, G., 1980, 'Integrated river basin planning in a holistic universe', *Natural Resources Journal*, **20** (4), pp. 787–806.

Selznick, P., 1966, *TVA and the Grass Roots*, Harper and Row, New York.

Sewell, W. R. D. and Barr, L. R., 1978, 'Water administration in England and Wales—impacts of reorganization', *Water Resources Bulletin*, **14** (2), pp. 337–48.

Sewell, W. R. D., Handmer, J. W. and Smith, D. I. (eds.), 1985, *Water Planning in Australia: from Myth to Reality*, Centre for Resource and Environmental Studies, Australian National University, Canberra.

Weber, E. W., 1964, 'Comprehensive river basin planning: development of a concept', *Journal of Soil and Water Conservation*, **19** (4), pp. 133–8.

Wengert, N., 1981, 'A critical review of the river basin as a focus for resources planning, development, and management' in R. M. North, L. B. Dworsky and D. J. Allee (eds.), *Unified River Basin Management*, American Water Resources Association, Minneapolis, pp. 9–27.

White, G. F., 1957, 'A perspective of river basin development', *Law and Contemporary Problems*, **22**, pp. 157–87.

CHAPTER 2

Integrated water management in the United States

Keith W. Muckleston

Introduction

Water is pervasive in biophysical subsystems, and its use is inextricably intertwined with many of society's efforts to enhance economic and social well-being. Thus, water is a substance of paramount ecological, economic and social importance. Interrelationships inherent in water use should encourage integrated management. In the United States as elsewhere, however, extensive implementation of integrated management has not been realized. Part of the challenge in the United States results from its large size and its subsequent hydrologic diversity, and part reflects its federal system whose 50 states exercise various degrees of governmental powers over land and water.

In this chapter the term 'integrated management' is used in a broad sense. It denotes the simultaneous management of two or more water-related phenomena. Such a process attempts to plan for and react positively to at least some of the myriad interdependencies within and between the human and biophysical subsystems related to water management actions.

The chapter is organized into four parts. First, consideration turns to the physical and ideological context within which integrated management has taken place. Second, the discussion focuses on co-ordinated management of water-related goods and services. Third, the river basin as a unit of management is presented. And fourth, the problem of integrating management of water and related lands is reviewed. A brief summary and observation on the future complete the chapter.

Context

Physical environment

The physical environment has played a mixed role in US water management approaches. The availability of moisture, reliability and amount of surface runoff and the size of drainage basins have all played a part. Relatively low levels of soil moisture and unreliable runoff encouraged the development of integrated management because these conditions require more active participation by the public sector. Through taxing powers, the public sector (usually the federal government) has been generally willing and able to provide a variety of water-related goods and services, most of which range from non-vendible to only partially marketable and therefore would not be provided by the private sector. Physical conditions in the West encouraged water resource development by the federal government, much of which became integrated based on the creation of large volumes of storage.

The conterminous United States may be divided into two macroregions: a relatively humid eastern half and a sub-humid to arid western half. This dichotomy is based on availability of surface moisture, reliability of runoff and availability of storage. Availability is derived by comparing average potential evapotranspiration and mean precipitation (Piper, 1965, Plate 2). Extensive areas in the western half yield less than 25 mm of runoff yearly and therefore do not support perennial streams (USGS, 1983, p. 16, Figure 5).

The relatively high reliability of runoff in the eastern half of the United States delayed the need for intensive management there. Reliability is determined by comparing the lowest annual runoff of the driest year out of 20 years with the mean annual runoff of a region in question (US Water Resources council, 1968, pp. 1–5). Assuming that regions with reliable runoff are those where the lowest year in 20 is 50 per cent or more of the mean, the east–west dichotomy is again apparent. Through much of the US water development history, unreliable runoff in the West was managed by the construction of large storage facilities by the federal government. Large storage capacity in turn provided the operational flexibility necessary for the integrated management of two or more water related outputs. An indicator of managerial flexibility is the percentage that existing reservoir capacity in a region constitutes of its annual renewable supply. Regions in the western half of the United States display markedly higher percentages than those in the east (USGS, 1983, pp. 32–3).

Ideologic context

Ideology has had a significant impact on the manner in which water resources are managed in the United States. Two major and conflicting precepts are present through much of the history of water management in the country: one would leave much of the water resource allocation to market forces; the other would rely more frequently on powers of government to regulate and/or allocate the production of many of the water-related goods and services. With the noteworthy exception of the Pro-

gressive period under Theodore Roosevelt, during most of this century
Republican administrations have preferred the first approach and Democratic administrations the latter. Federal management of multipurpose
projects often resulted in approaches that attempted to co-ordinate, or
integrate operations.

Nineteenth century. During the first half of the nineteenth century,
integrated water management was largely absent. The predominant
ideology regarding the role of government in water management was one
of rugged individualism. Water management units—farmers, municipalities, mill owners, barge operators, etc.—were on their own. Most
management actions would have fallen within White's first strategy: they
were undertaken by private entrepreneurs, pursuing a single purpose, who
used the limited structural means available and gave little attention to the
impacts of their actions on other water users (White, 1969, pp. 15–33). The
federal government was rarely involved except with navigational improvements. Some of the eastern states became active in expensive canal
construction prior to the ascendancy of railway transport, while some cities
embarked on 'municipal socialism' by organizing public water supply
systems (Mumford, 1961, p. 476).

During the last quarter of the century, the federal government became
increasingly involved in the settlement of the West. The involvement
progressed from encouragement to greater levels of actual assistance.
Under the Desert Land Act of 1877 federal lands were sold to settlers who
would irrigate them within three years. In 1890, statutes reserved canal
rights-of-way west of the 100th meridian for the federal government that
might later be used for irrigation (Viessman and Welty, 1985, p. 31). In
1894 the Carey Act attempted, largely without success, to cede up to
465,000 ha of land to any western state that agreed to arrange for its
irrigation. The increasing level of federal involvement in the settlement of
the West culminated in the preparation for passage of the Reclamation
Act.

The Progressive period, 1900–21. During this era, much of the ground-
work was laid for later efforts to manage water and related resources in an
integrated manner. A number of ideologies affecting the management of
land and water during this period have been identified by Viessman and
Welty (1985, pp. 31–5). These include conservation, support for small
independent entrepreneurs and encouragement of equal opportunities, all
to be supported by strong government action. The idea of the multipurpose
use of the newly acquired national forests was expanded to include
multiple uses of water and the interconnections between lands and waters
within a drainage basin. Recommendations of three official study commissions between 1908 and 1912 strengthened the argument that since

natural resources were bound in a web of interrelationships, their management should reflect that through integrated approaches.

Ideology also affected private development of hydropower sites during this period. Between 1884 and the passage of the 1906 General Dam Act, private development of hydropower sites had received *laissez-faire* treatment through a number of Congressional authorizations allowing construction of privately owned power installations on navigable waters (Price, 1971, pp. 17–19). Five vetoes of subsequent bills by President Roosevelt then sharply reduced development through much of the remainder of the Progressive period. The Federal Water Power Act of 1920 (FWPA) sought to strike a balance between the public and private sectors by allowing private development with the requirement for public licensing and regulation. This shift back towards private development, albeit with public regulation, signalled the beginning of a new twelve-year period of conservative Republican administrations.

Post First World War period. This period must be considered on two levels regarding the effects of ideology on integrated water management. At the highest and rather abstract level of party platforms, the 'return to normalcy' which was presumably taking place meant that private enterprise was allowed to flourish with a minimum of governmental competition (Viessman and Welty, 1985, p. 35). This would have retarded integrated management, but strong undercurrents of Progressive ideologies carried over from the earlier period. Congress increased authority of the Corps of Engineers (CE) to carry out basin-wide multipurpose plans (the 308 Reports). The Boulder Canyon Act was passed in 1928, which was the first time that Congress had authorized explicit approval of multiple outputs from one large structure (White, 1969, p. 35).

At the end of the period, widespread economic chaos set the stage for reactive policies of the next era, many of which led to increased levels of integrated water resources management.

New Deal period. This era lasted from the inauguration of President Franklin D. Roosevelt in 1933 to the Congressional abolition of his Natural Resources Planning Board in 1943. Many of the natural resource management decisions of this period were designed to rescue the economy from the Great Depression and drought through construction of large public works. Ideologies of the Progressive period re-emerged and were vigorously pursued. Examples include expansion of family irrigated farms through increased Bureau of Reclamation (Bu Rec) activity; encouragement of public power through exercise of the Preference Clause (Viessman and Welty, 1985, pp. 38–9); and formation of the Tennessee Valley Authority (TVA), the crown jewel in the New Deal's plans for comprehensive natural resource utilization.

Centralized planning under the aegis of the executive branch was carried out by the National Resources Planning Board (NRPB) and its predecessors. An unprecedented and controversial step by the NRPB was its consideration of social impacts stemming from land and water developments. The close of this era was signalled by Congressional abolition of the NRPB which had attempted to co-ordinate planning, horizontally among government agencies and vertically among federal, state and local actors. The NRPB's function as filter between Congress and the federal construction agencies—and also between Congress and potential project beneficiaries—proved unpopular, leading to its demise (White, 1969, pp. 70–1).

Post Second World War construction boom. This period was characterized by extensive public development of water projects, many of which were multipurpose and therefore included a degree of integration between water-related goods and services. Integration between agencies and levels of government declined after abolition of the NRPB in 1943 (Viessman and Welty, 1985, pp. 39–42). The term 'basin planning' was widely used but less frequently employed as development tended to proceed on a project-by-project basis. This may be attributed to the non-reimbursability of most water-related outputs, the constituency orientation of Congress, the informal coalition between Western irrigation interests and southern flood control and navigation interests, and the absence of an effective co-ordinating mechanism. In 1944 Congress set the tone for much of the period by the passage of the Pick–Sloan Bill which authorized extensive construction within the Missouri Basin, an action which is widely considered the epitome of uncoordinated development through large-project-oriented construction.

Political ideologies of both major political parties influenced integration of water management during this period. During the years 1943 to 1952 and 1961 to 1965, Democratic administrations continued to favour public works as one means of enhancing the economic and social well-being of the country, but as noted above, the lack of effective co-ordinating mechanisms often restricted integration to several water-related outputs from large individual structures. In contrast, the Republican administration of 1953–61 actively discouraged the construction of large federal projects (Eisenhower's 'no-new-starts' policy), while encouraging non-federal development of hydropower, the only lucrative output of water projects (Price, 1971, pp. 32–4).

Throughout the latter part of this period, currents counter to water management by means of large structures became increasingly strong. Growing environmental deterioration, greater affluence and more leisure time were contributing to the growth of the environmental movement with its concerns for water quality, non-structural approaches to management and leisure time/aesthetic uses of water.

Deteriorating water quality was one of the principal areas of concern. Between 1948 and 1961, several tentative steps were taken towards greater federal involvement in matters of water quality. Although Congress could have used the Commerce Clause 'to pre-empt' water pollution control, interests favouring more federal action continued to have difficulty because water quality problems were widely perceived as a matter for state and local governments (Hines, 1971, pp. 456–75), a view espoused as policy when the Republicans were in power.

Ascendancy of environmentalism and incipient federal disengagement. During this era (1961–81) there were significant changes at the national level in policy and management of water resources. The rapid rise of the environmental movement during the 1960s resulted in an outpouring of actions by the legislative and executive branches designed to protect and enhance the environment. In addition, less evident but persistent signals indicated that the federal government was beginning to disengage from its well established role as builder of large structural water projects that provided many non-reimbursable benefits. The ideological dichotomy of earlier periods between Democratic and Republican administrations was muted as both environmentalism and initial moves toward disengagement continued across administrations.

Environmentally oriented actions affected integrated water management in a variety of ways. Because many of these actions emphasized protection and preservation, they were the antithesis of multipurpose land and water use which had stressed the production of traditional/utilitarian outputs often at the expense of leisure time/aesthetic values. For example, the Wild and Scenic Rivers Act of 1968 (PL 90-542) preserved the natural qualities of eight river segments and provided a mechanism by which many more have been added to the system. Another example is the voluminous and multifaceted Water Quality Act of 1972 (PL 92-500). It contained provisions, which if implemented, would have eliminated the waste carriage function of water by setting a 'no discharge' goal for 1985. This unrealistic goal was criticized by the National Water Commission (NWC) which preferred instead provisions of the Water Quality Act of 1965 (PL 89-234), because it had set standards for receiving waters thereby assuring multiple instream uses of some rivers while preserving high quality waters for others (NWC, 1973, pp. 69–71).

On the other hand, some actions of this period required increased attention to interdependencies. Section 208 of PL 92-500 encouraged countering nonpoint sources of pollution through land use management. In 1973 the Water Resources Council (WRC) promulgated the Principles and Standards which required federal agencies planning project construction to evaluate trade-offs between national economic development and environmental quality, while also considering regional development and social well-being. The National Environmental Policy Act of 1969 (NEPA) required federal agencies to recognize interrelationships between water projects and related biophysical phenomena. However imperfect some of the resulting project environmental impact statements may have been, NEPA contributed to a more holistic approach to water development.

The second major development of this era was the movement toward federal disengagement from its traditional role as provider of large structural projects. This new direction was obscured by the commencement of some large federal water projects. Projects the magnitude of the Central Arizona Project and the Tennessee–Tombigbee navigation project, suggested at the time that it was business as usual despite the strong environmental movement. However, these and other large projects had been in the planning stages for years. Due to the protracted gestation period of major projects, they came to term during this era.

Passage of a major policy act in 1965 also contributed to the perception that strong federal presence in the water development field would continue. The Water Resources Planning Act (WRPA) was designed to improve co-ordination among federal agencies and improve project efficiency. However, the WRPA (PL 89-80) also contributed to disengagement because it contained provisions to devolve upon states more planning responsibilities. As Gregg (1989, pp. 14–16) notes, the WRPA was unsuccessful for several reasons: the major national concern over water had shifted to quality (which it could not effectively address), traditional federal projects had become symbols of inefficiency and environmental degradation, the Office of Management and Budget remained hostile to spending for traditional water projects and increasing attention was being paid to non-structural approaches to water resources management.

Another aspect of disengagement was the growing attention to increased cost-sharing to be borne by non-federal levels of government and other project beneficiaries. This eroded support for traditional federal projects of former backers by undermining the symbiotic relationship between special interest groups, federal construction agencies and Congress. The rapidly escalating expense of proposed projects made prospects of increased cost-sharing percentages even more unpalatable. ·

Devolution and counterenvironmentalism, 1981–9. Unlike the previous era, ideology appeared to be central in a number of decisions affecting water management. The Reagan administration was markedly influenced by archconservative ideologues whose opinions on the relative priorities of military and domestic spending, the efficacy of market forces *vis-à-vis* government institutions and the environmental movement affected water resource use and development.

Fiscal austerity for the domestic sector resulted in a marked diminution of projects constructed and/or supported by the federal government. For example, in 1985 the CE reported that of the 106 ongoing construction projects only six had begun after 1979 (Wall and Schilling, 1985, p. 22). It is noteworthy that the decrease took place even though the administration had vitiated environmental regulations and simplified project planning procedures in order to facilitate project construction.

Integrated water management was affected adversely as the mechanisms to co-ordinate, provide research and plan were either dismantled or eviscerated. Most of the already weakened institutions set up under the WRPA were summarily eliminated. Within the first year the WRC and all of the river basin commissions co-ordinated under it were abolished in addition to funds for state water planning assistance. In the following year, the termination of the Office of Water Resources and Technology, which had been responsible for administering funds from the Water Resources Research Act, reduced water resources research, particularly by universities. The Principles and Standards were scrapped in favour of the Principles and Guidelines (P&G). This action reduced to four the federal agencies required to use the P&G during project planning and largely eliminated environmental quality, regional development and social well-being from consideration in favour of national economic development (Viessman and Welty, 1985, pp. 123–6). The need to re-establish co-ordination at the federal level and between levels of government was recommended, however, even by those accepting the fact that federal programmes 'had matured' from a stage of development to one of operation and maintenance (Schilling et al., 1987, p. xiii).

At the state level, planning and co-ordination of water resource activities have progressed unevenly, as devolution of general responsibilities for many water-related matters has, perversely, not been accompanied by federal funds to carry out the new responsibilities (Clark, 1985, pp. 12–13). A related facet of devolution is the continued modification of cost-sharing formulas. Passage in 1986 of the first major omnibus water development bill (PL 99-662) in 16 years shifted significantly more costs of federal projects to non-federal levels of government and project beneficiaries. The lengthy hiatus between passage of major omnibus bills reflects in part the difficulty of reaching agreement on the percentage and nature of cost-sharing. In another state-related matter, it is unclear the extent to which the Reagan administration's ideology and devolutionary actions affected the many and varied attempts in the West to use prices and/or market-like mechanisms to reallocate some water from agriculture to higher value uses. Although attempts to circumvent the rigidities of the prior appropriation system were present in the earlier era, they became significantly more common in the 1980s.

Counterenvironmentalism, as devolution and dismemberment of institutions, was facilitated by fiscal austerity for domestic programmes. A case in point is the 1986 pocket veto of the 20 billion dollar Clean Water Bill (S 1128) which had been passed unanimously by both chambers. At other times the administration used a combination of ideology and fiscal austerity to counter the environmental movement. The effectiveness of EPA was diminished by underfunding and by demoralization of its professional staff through appointment of ideologues to positions of leadership within the organization. Other actions were undertaken on ideological grounds. For example, the Heritage, Conservation and Recreation Service (formerly Bureau of Outdoor Recreation) was abolished and an attempt was made to eliminate rivers being considered for inclusion under the Wild and Scenic Rivers System because they contained private lands or if developers had objected (Palmer, 1986, pp. 156–7).

Integrated management of water-derived services

Introduction

This section focuses on the integrated management of water-derived services. More than a century ago some of the individual projects on the lower Mississippi included navigation, flood control and drainage. Since then techniques of integration have become more sophisticated, allowing not only the co-ordinated management of a greater number of water-derived services but also integration between projects within a drainage basin.

If the term 'integrated management' denotes '. . . the simultaneous consideration of formerly separate resource development problems' (Wescoat, 1984, pp. 7–8), then this management technique has indeed a long and extensive history in the United States; but if the definition describes 'the management of a system of complementary facilities under single overall control . . . in a manner which maximizes the combined net benefits from the operations of individual reservoirs and other facilities' (Ackerman, 1960, p. 6), examples are less frequent and of more recent origin. In the late nineteenth and early twentieth centuries integrated management of multiple water-derived services resembled the former definition; from the 1930s until the 1960s it moved towards the latter, although 'maximization' of combined outputs was at best an illusive goal. During the 1970s and especially the 1980s, single purposes appeared to gain in popularity while integrative mechanisms were de-emphasized.

Federal management

Because few water-derived services are readily marketable, the public sector has been largely responsible for the integrated management of water-derived services. Taxing powers financed the large, expensive structures and extensive earth-moving operations that are often necessary to realize scale economies. With relatively few exceptions, the principal actors have been the CE, Bu Rec, TVA, and since the 1954 passage of the Watershed Protection and Flood Prevention Act (PL 566), the Soil Conservation Service (SCS) whose smaller projects also produce multiple outputs. The dominant approach to the integrated management of water-derived services fits White's third strategy: public ownership, producing multiple outputs and using structural means (White, 1969, pp. 13, 34–46). This approach is also characterized by the management of supply with little effort given to the modification of demand for the services in question.

Most of the integrated water management techniques gained at least partial legitimization through Acts of Congress. While the 1879 Act which created the Mississippi River Commission recognized the interrelationships of navigation, flood control and bank protection, numerous additional acts through the subsequent five decades refined and extended techniques of managing multiple water-derived outputs in an integrated manner, culminating in the TVA Act of 1933 (Ackerman and Loef, 1959, pp. 508–9). During the first few decades of integrated water management, four water-derived services received most of the attention. In chronological order they included: inland water navigation, flood control, irrigation and hydroelectric generation. With the exception of irrigation, all are instream uses which are often complementary but sometimes substitutable; therefore management to attain multiple use may cause conflict between their respective user groups. When storage projects are employed, irrigation is often a complementary use. During the 1950s and 1960s integrated management of traditional utilitarian outputs added municipal and industrial water use as well as leisure time/aesthetic outputs, which include recreation and the mitigation and enhancement of fish and/or wildlife. With the addition of each water-derived service the potential for conflict between project beneficiaries rises, thereby increasing the complexity of management (Sewell, 1976, pp. 794–5).

The role of large dams. Economical construction of large dams was the *sine qua non* of modern forms of integrated multipurpose water management: the ability to store unprecedented volumes of water behind huge structures was the key to more sophisticated integrated management. Hoover Dam, authorized by the Boulder Canyon Act of 1928, is generally recognized as the prototype of this type of structure.

Large dams have been most widely used in the West where, for example, the ratio of normal reservoir storage to annual average runoff is 4.22 in the Colorado Basin, 1.89 in the Rio Grande Basin, and 1.12 in the Missouri Basin; whereas in the eastern half of the United States, these ratios range from 0.95 to 0.26 (Foxworthy and Moody, 1985, pp. 63–4). The expense of providing storage rose rapidly after the best storage sites were utilized. If the real cost of storage provision is assumed to be the volume of reservoir capacity created per unit volume of dam (hm^3 storage water per m^3 of dam material), then between the pre-1921 period and 1970 the storage created per unit of dam material decreased by over 97 per cent (USGS, 1983, p. 33).

Three developments may be attributed to the economic construction of large dams during the early part of this century (Ackerman and Loef, 1959, pp. 219–34, 259–66). First, the revolutionary development of earth-moving equipment particularly facilitated the construction of large earth-fill dams and also aided in reservoir site clearance and tunnelling. Second, innovations in concrete production and construction techniques allowed dams of unprecedented height to be constructed. Finally, marked improvements in the generation and transmission of hydroelectric energy allowed public management entities to subsidize other water-related outputs with the sale of this readily vendible output, the demand for which grew astronomically during this century.

The role of non-federal producers of hydropower

Non-federal power producers also practice some integrated water management. Although these actors now assume significantly more responsibility for other water-related goods and services than they did early in the century, the degree of integration usually does not approach that employed by the major federal entities. Since 1920 most of the impetus for integrating other water-related goods and services into management decisions has been achieved through the licensing and regulation of non-federal hydropower producers by the Federal Power Commission (FPC), and since 1977, by its successor, the Federal Energy Regulatory Commission (FERC). These regulatory actors were created under provisions of the 1920 FWPA and its amendments.

Concern by non-federal hydropower producers for provision of other water-derived services may be considered within three categories: (1) mitigation of adverse effects resulting from hydropower development; (2) provision of water-related goods and services downstream from a storage facility created as part of the project; and (3) on-site provision of another water use at the project reservoir. Under the first category, navigation of inland waterways has been protected since early in the century through construction of locks (or the provision therefor), maintenance of channel depth and operation of turbines so as not to interfere with navigation. Decades later the Fish and Wildlife Coordination Act of 1934 and its amendments (1946, 1958) progressively strengthened measures to mitigate adverse effects of dam construction on fish and wildlife, with special consideration given to anadromous fish (US Senate, 1960, p. 16). Mitigating damage to the environment was substantially improved by passage of NEPA in 1969.

The second category pertains to hydropower projects that have storage capacity (as opposed to pondage behind run-of-river dams). In these instances obtaining a license may be conditioned upon use of some of the storage capacity for non-power outputs downstream, which usually reduces the value of hydropower output at the dam site and at any downstream facilities the licensee may operate. For example, in 1955 the FPC required Idaho Power Company to provide approximately 0.612 km^3 of storage space for flood control by 1 March each year at Brownlee Dam on the Snake River.

Slack water recreation is the most common output of the third category of co-ordinated water management by non-federal hydropower producers. Slack water recreational opportunities may be created together with the impoundment found behind dams built for hydropower generation. Enhancement of public relations encouraged some utilities to provide voluntarily recreational opportunities, but the first major impetus came from the FWPA. In the original act (1920) recreation was not specifically mentioned in language directing comprehensive planning. In the 1935 amendments recreation was specifically included under Section 10(a), but in 1962 the influential Outdoor Recreation Resources Review Commission (ORRRC) concluded that FPC's treatment of recreation and access to reservoirs had been '. . . generally passive and permissive' (ORRRC, 1962, pp. 13–15). Since 1965 a general policy to correct this criticism has been in force.

Since 1970 relicensing of non-federal hydropower projects has provided the FPC and its successor with ample opportunity to impose new conditions on the producers. The timing of this development results from the passage of the FWPA in 1920, which broke a long standing backlog of pending non-federal hydropower projects, and from the 50-year licence period. Rapid development of non-federal hydropower in the decade after passage of the FWPA meant that during the 1970s a large number of non-federal producers had to apply for relicensing. Investor-owned utilities have been particularly anxious to demonstrate an uncharacteristic concern for water-related goods and services other than hydropower: under Section 14 of the FWPA the federal government may take over an entire project at the end of the 50-year period, and may do this by payment only of net investment and severance damages (Price, 1971, pp. 46–8). Therefore, prior to the relicensing process detailed plans for the expansion of recreational opportunities, fish and wildlife enhancement and environmental protection features are submitted by the non-federal licensees to build a strong case against take-over.

Catchments as units of management

Introduction

Various approaches to basin management have been used in the United States over the last several decades. In general they have had limited success and may be characterized as *ad hoc*, transient and/or experimental

(Hart, 1971, p. 1; Derthick, 1974, p. 1). The advantages of using catch-ments as units of management have been advocated by the intellectual and planning communities since the nineteenth century (Wengert, 1980, p. 11). Supporters argue that this approach would mitigate some of the basic problems facing co-ordinated management by spatially matching the units of supply (the catchments) with units of jurisdiction. The major impedi-ment to this approach appears to be the established actors in the federal system: federal agencies and state governments view regional management schemes—whether based on surface hydrography or some other phen-omenon—as a threat to their legitimate roles. They therefore usually resist the formation and effective functioning of regional management units.

The large size of many river basins in the United States presents an additional problem to co-ordinated basin-wide management because two or more states have all or parts of their territories within common catch-ment borders. A map of interstate river basins in the conterminous United States reveals few significant intrastate rivers (White, 1957, fig. 2, p. 196). A limited number of intrastate basin management units are present but most do not operate with a high degree of integration (NWC, 1973, pp. 414–16; Canter and Cristie, 1984, pp. 32–3).

Approaches used

Interstate compacts. By using interstate compacts, states can to a limited degree manage interstate waters without federal participation. This approach must be approved by Congress and has been used since the earliest years of the Republic, although much more frequently during the last 50 years. Three general water-related issues are dealt with by com-pacts: water allocation, pollution control and flood control (NWC, 1973, pp. 415–21). Interstate compacts have not provided comprehensive water management because they are limited to a few water-related issues and generally cannot influence water management by federal agencies within the basin covered by compact.

Interstate compacts with federal participation. When the federal govern-ment is involved in interstate management of water resources, the various approaches used may be divided into two types. The first can only plan and co-ordinate. In addition to planning and co-ordination the second may do any of the following: regulate, construct and operate projects. The first type has been represented by a Federal Interagency Committee (FIAC) structure and Title II River Basin Commissions (RBC). Since the 1940s, FIACs have operated at various times over some large river basins of the United States, including the Missouri, Columbia, Pacific Southwest Basins, Arkansas-White-Red and New York-New England. Consisting of federal agencies active in water development in the respective drainage basins, these informal organizations function as forums to allow better co-ordination among federal agencies and with state level interests. No staffing and little input from states limit the effectiveness of these organizations, whose plans are usually weakly co-ordinated assemblages of individual agency plans.

The RBCs (1968–81) provided more comprehensive plans, involved the states to a greater degree and had marginally improved staffing arrangements. The RBCs operated under authority derived from Title II of the WRPA passed in 1965. Between the late 1960s and middle 1970s approximately 30 states agreed to form seven such RBCs which covered almost all of the northern third of the conterminous United States. The plans produced by the RBCs—the comprehensive, co-ordinated, joint plans for water and related land resources—were intended to function as guides for future Congressional authorization of water projects. Weaknesses in the RBC approach included lack of full acceptance by the traditional water development establishment, inability to address issues within the drainage basins due to requirement for consensus, uncertainty about the WRC–RBC relationship and underfunding of planning operations. After abolition of the RBCs in 1981, FIACs have been re-established in some of the areas they once served.

The second type of interstate, catchment-based organizations are those with powers beyond co-ordination and planning. Only three such organizations exist in the United States: two federal–interstate compacts, in the Delaware and Susquehanna Basins; and one federal corporation, the TVA. The federal–interstate compacts are relatively new actors dealing with interstate water problems. The Delaware River Basin Commission (DRBC) was authorized in 1961 and the Susquehanna River Basin Commission nine years later. This type of commission consists of a representative from each participating state and one federal member who represents the Secretary of the Interior. Content of the comprehensive plans reflects majority vote, thereby overcoming the inaction resulting from the RBC's consensus requirements. Escape clauses allow the federal government to refuse to comply with management actions with which it disagrees. It is noteworthy, however, that legislators felt this would be seldom used and that relatively harmonious federal–state relationships have prevailed within the federal–interstate commissions.

The federal–interstate commissions are authorized to exercise a broad range of water management techniques. The DRBC, for example, may regulate water use through licensing, construct and operate projects, and raise revenue for operations (Derthick, 1974, pp. 46–75). The NWC preferred the federal–interstate compact as a mechanism to resolve interstate water management issues; it is recommended that this approach be utilized widely, particularly in the West where the federal government is the dominant landowner and water master (NWC, 1973, p. 424).

The TVA is the most renowned catchment-based water management organization in the United States. Although the TVA manages a catchment area of only about one per cent of the country's area and even though this type of organization has not been replicated in the United States, it continues to be offered as a model of river basin management for other countries (Teclaff, 1967, pp. 132–43).

Created at the zenith of the New Deal period, TVA is a federal corporation empowered to manage comprehensively a number of water-related goods and services (Derthick, 1974, pp. 18–45). In addition, it has responsibility for certain land based resources and is even involved with some socio-economic and educational activities. Principal water-related outputs under TVA's jurisdiction include navigation, flood control and the production and wholesaling of electrical energy from hydro and thermal plants. These activities made TVA a competitor to well established actors in the basin, particularly the CE. In addition, TVA's energy-related activities, which included strong support for public utilities both within and outside the catchment, sharpened widespread conflict between public and investor-owned utilities.

Management activities on water-related lands include floodplain management, in which TVA was a pioneer. TVA activities on land-related resources include aspects of reforestation and improvement of marginal agricultural lands, traditionally the concern of the US Forest Service and SCS, respectively. The production and sale of inexpensive fertilizer to stimulate agricultural productivity and munitions production are activities carried out by no other US water management agency. TVA also assists industrial development by improving inland waterways markedly.

Inability to replicate the TVA experiment in other parts of the country largely resulted from TVA's conflict with state governments and federal agencies which resented the usurpation of their powers·by this regional organization. Moreover, much improved economic conditions removed the necessity to experiment with management approaches that many considered radical. The NWC noted that replication of the TVA model was neither advisable nor feasible (NWC, 1973, p. 427); while even the generally bland, neutral conclusions of the WRC rejected expansion of the TVA model (USWRC, 1967, p. 8).

Evolving management strategies in the Columbia basin

Since 1980 a prototype interstate compact has been in operation. Passage of the Pacific Northwest Power Planning and Conservation Act (PL 96-501) enabled the states of Washington, Oregon, Idaho and Montana to create the Northwest Power Planning Council (NPPC). This regional council is charged with formulating two detailed plans: one that seeks to ensure the region has an adequate electrical power supply and the other that is designed to protect, mitigate and enhance fish and wildlife—

especially salmonids in the Columbia basin. The NPPC is an innovative and controversial actor. It has no federal representative and consists of eight gubernatorial appointees whose plans for fish and wildlife and electrical energy significantly impact some of the plans and operations of long established, mission-oriented entities, most of which are at the federal level. Indeed the unusual requirement that some of the plans and operations of federal agencies be substantially in compliance with the plans of a non-federal interstate compact commission is requiring years of difficult adjustments to implement fully. The NPPC, its staff and operations are financed through the sale of electrical energy by the Bonneville Power Administration (BPA). Thus consumers of federally generated hydropower are contributing to the restoration of fish and wildlife.

Despite the resistance from established entities, the NPPC has made considerable progress toward meeting its goals, particularly in regard to rebuilding stocks of salmonids. An innovative management tool is the water budget which utilizes 5.69 km^3 of storage releases in the spring to speed migration of juvenile salmonids (smolts) to the ocean. Trade-offs result because storage water for this artificial freshet was formerly reserved for generation of firm energy in the autumn and winter, the period of highest demand for electrical energy in the region. The NPPC programme also requires accelerated completion of costly bypass facilities at dams owned by public utilities and the CE. Until completion of the bypass systems during the 1990s, controversial spill procedures are called for at the 13 run-of-river dams on the mid and lower Columbia and lower Snake. Spill is intended to provide an alternative and less harmful means of dam passage than the route via penstocks and turbines. Voluntary spill means energy is foregone for decreased smolt mortality. Limited spill took place for several years but was vigorously opposed by hydropower interests until 1989 when a negotiated agreement between most parties was finally concluded.

Integration of water and associated land resources

Introduction

The integrated management of water uses with those of related lands would seem only natural. Spatial, temporal, quantitative and qualitative attributes of water can affect associated lands in many ways; conversely, natural characteristics of lands and their uses can affect the same four attributes of water. Yet, despite the many linkages between water and associated lands, their integrated management in the United States is the exception rather than the rule. Although a number of efforts to address the interrelationships of land and water have been made, their overall impact

pales relative to those of the more numerous and better funded programmes which focus on either land or water.

The lack of integrated management may stem from some significant
differences between the two phenomena, both real and perceived. Land is
owned in fee simple, while usufructuary rights are the rule for water. Land
is relatively fixed in space, whereas the mobility of water through and
across political entities and parcels of privately owned land is one of its
fundamental characteristics. Few question that land has monetary value,
while water and/or some of its derived goods and services are not infrequently perceived to be free goods, akin to air or sunshine. Perhaps
because of these distinctions, different agencies often have management
responsibilities for physically interconnected land and water resources.
This fragmentation fosters uncoordinated and sometimes conflicting
patterns of resource use.

Flood-hazard mitigation

Flood-hazard mitigation, which considers both land and water, provides an
example of efforts to bring an integrated approach to water-related problems. It is contrasted with the earlier 'flood control' approach which
concentrated on controlling flood waters while largely neglecting flood-
prone lands. Serious national attention to an holistic approach to addressing flood hazards began less than three decades ago but now has thousands
of participating communities. The problems and issues of implementing
the approach illustrate many of those associated with managing land and
water in an integrated manner.

Background. Flood hazards reflect the interrelationship of land and
water. The net benefits which are derived from using floodplains are
eventually reduced, or even eliminated, when rivers rise to reclaim
temporarily their floodplains. Clearly, an integrated approach appears
warranted. But, until approximately 25 years ago most efforts at flood-
hazard reduction were focused on controlling flood waters. This structural
approach started on a large scale after passage of the Flood Control Acts of
1936 and 1938, when Congress accepted flood problems as a national
responsibility. Despite large expenditure on structures 'to keep water away
from man', flood damages continued to rise, in part because little attention
was paid to land use in flood-prone areas.

A significant juncture was reached in the mid-1960s when a small task
force of experts chaired by Gilbert F. White produced a report entitled *A
National Program for Managing Flood Losses*. Promulgated as House
Document 465, it is widely perceived to be both the turning-point towards,
and blueprint of, a more balanced and efficacious approach to flood-hazard
mitigation. HD 465 also stimulated Congress to focus attention on the
problem of rising flood losses.

The National Flood Insurance Program (NFIP).   Two years after promulgation of HD 465, the National Flood Insurance Act (NFIA) of 1968
(PL 90-448) was passed. Flood insurance became available in those communities that agreed to adopt and enforce floodplain management standards promulgated by a federal agency. Over the last two decades the
NFIP has undergone several modifications, including a change in management agencies as well as moving from a voluntary programme to one that
became practically mandatory after passage of the Flood Disaster Protection Act of 1973 (PL 93-234) (Muckleston, 1976, p. 56). Whereas in
1971 only 158 communities were participating in the NFIP, by 1979 almost
17,000 were (Platt, 1986, p. 56). A decade later there were approximately
20,000 communities practising land use regulations under the NFIP. Value
of property insured under the NFIP also grew rapidly: from $1.1 billion in
1971 to $162 billion in early 1988 (Platt, 1986, p. 56; USGAO, 1988, p. 2).
    Congress provided another means of managing flood-prone lands under
the NFIP. Section 1362 of the NFIA allows lands with severe flood
problems to be purchased by the federal government for transfer to more
harmonious uses. The Federal Emergency Management Agency (FEMA)
began to utilize Section 1362 in 1980 after Congress finally appropriated
funds for that purpose. Under this section FEMA purchases lands with
repeated flood problems for transfer to state or local governments. Non-
federal governments accepting title to such lands must agree to manage
them for open space or other non-development purposes (USGAO, 1988,
p. 14). During the first eight years of this approach almost 1,300 parcels of
property were approved for purchase with an estimated cost of $36 million
(USGAO, 1988, p. 17).
    Despite the widespread participation in the NFIP and the greater
attention it brought to integrating land and water management, it should
not be considered an unqualified success: flood-related damages continue
to rise; the NFIP as presently constituted is probably unsuited to mitigate
flood hazards in coastal areas, where the majority of insurance policies are
now found; it has run large deficits for most of the 1980s; and inability to
monitor adequately enforcement of land use regulations has led to uneven
implementation. While the NFIP is unquestionably more comprehensive
than earlier approaches to flood-hazard mitigation, it only addresses four
of the multifaceted aspects of the problem: land use management, flood-
proofing, insurance and land acquisition (Platt, 1986, p. 59). A more
comprehensive approach is needed that would use the many tools available, including more comprehensive co-ordination of land and water
resources.

A Unified National Program for Floodplain Management (The Program).
The plan for such an approach was promulgated by the WRC in 1979 and
reissued by FEMA seven years later. The Program had its origins in NFIA,
which called for its delivery to Congress in two years. The dilatory delivery
of this document eleven years later reflects the inherent difficulty that

mission-oriented agencies have in agreeing to a unified approach—even at a theoretical level. Since the WRC had been assigned the task of producing the document, the consensus requirement among participating agencies made development of a meaningful plan protracted and difficult. A version of the Program was first adopted by the WRC in 1976, but was not sent to Congress for another three years as the Carter administration required incorporation of floodplain natural values into the document.

The Program calls for the combined use of three major strategies: the modification (reduction) of (1) the susceptibility to flood hazards, (2) the floodwaters themselves and (3) the impact of flooding (FEMA, 1986, pp. IV 1–16). While the Program recognizes that the relative significance of the three strategies should vary on a case-by-case basis, reducing susceptibility is emphasized. The major part of this strategy involves various types of land use management: floodplain regulations, zoning, subdivision regulations, codes for building and housing construction, etc. The Program recommends that when flood crest reduction is employed, pre-existing floodplain regulations and flood preparedness plans be in place prior to construction. Thus, the thrust of the Program is a marked departure from the structural approach that was relied upon after the Flood Control Act of 1936. The dominant theme of the Program appears to be that the emphasis should be on keeping inappropriate development away from the water rather than water away from development.

Utilizing the natural values of floodplains is also offered as a management tool in flood-hazard reduction. Land and water relationships are an integral part of this strategy. The value of floodplains in flood storage and conveyance, as well as for water quality maintenance and groundwater recharge are noted with directions on how these may be maintained. The inclusion of natural values reflected the Carter administration's concern with environmental issues. Given the Reagan administration's prevailing philosophy on environmental matters, it is noteworthy that the chapter on natural values was retained in the 1986 edition of the Program.

Implementation of the Program remains problematic in the face of formidable institutional obstacles. Although the techniques and many of the authorities for actions are present, overcoming fragmentation would require unprecedented co-ordination. Three major types of co-ordination are needed: functional, vertical and horizontal. Regarding the first type, at the federal level alone 27 agencies acting under nine different legislative programmes are involved in various aspects of flood-hazard mitigation. Given the dismemberment of co-ordinative machinery by the Reagan administration, it appears doubtful that FEMA's Interagency Task Force on Floodplain Management will be more successful than the WRC was in overcoming problems of functional co-ordination at the federal level. In addition, fragmentation is present at the state level.

Vertical integration between levels of government may present an even more difficult challenge. Whereas the leadership for unified management comes from the federal government, police powers over land use rest with non-federal levels. The federal level can encourage co-ordination through grants-in-aid, technical assistance, provision of data and other relevant information; and it can mandate selected actions as through the Flood Disaster Protection Act of 1973. But, federal levels of funding for these purposes, while inadequate during the 1970s, have declined further in this decade.

Horizontal co-ordination is also necessary to manage hydrologic inter-dependencies occurring between states, and especially between local governments. Land management actions by local governments, or actors subject to their regulation, may alter flood hazards in other jurisdictions that are upstream, downstream, or cross-stream. Flood-related inter-dependencies at the interstate level are less apparent and can be addressed by interstate compact, although it is infrequently done.

The Balkanized approach to flood-hazard mitigation may be expected to continue until improved co-ordination takes place among functional actors, different levels of government, and subnational levels of government. While the Program offers a conceptual framework for efficacious action, it remains hortatory. It is doubtful that exhortations from the federal level will be followed with alacrity by subnational levels of government.

Summary

The integrated management of water and related lands in the United States has been affected by a variety of phenomena: biophysical characteristics, ideological positions on the role of government in resource use, as well as technological and institutional innovations.

For more than a century the federal government fostered most of the integrated development based on the desire to produce non-vendible outputs from water and related lands to attain broad social goals. During the last two decades water development responsibilities have increasingly fallen to subnational levels of government which have not yet filled the vacuum created by federal withdrawal.

Integration of land and water has also been attempted through util-ization of drainage basins as units of management and by more output specific programmes such as flood-hazard mitigation. Established institu-tions within the federal system have generally vitiated the effectiveness of basin-oriented management units, while the states' powers over land use have sometimes adversely affected both catchment organizations and specific programmes.

As the United States enters the 1990s it appears that non-federal levels of government, or actors operating under their institutions, must take the lead in fostering integrated management. The prognosis is by no means clear.

Abbreviations used

Bu Rec	Bureau of Reclamation
CE	US Army Corps of Engineers
DRBC	Delaware River Basin Commission
EPA	Environmental Protection Agency
FEMA	Federal Emergency Management Agency
FERC	Federal Energy Regulatory Commission
FIAC	Federal Interagency Committee
FPC	Federal Power Commission
FWPA	Federal Water Power Act
NEPA	National Environmental Protection Act
NFIA	National Flood Insurance Act
NFIP	National Flood Insurance Program
NPPC	Northwest Power Planning Council
NRPB	National Resources Planning Board
NWC	National Water Commission
ORRRC	Outdoor Recreation Resources Review Commission
RBC	Title II River Basin Commission
TVA	Tennessee Valley Authority
USGAO	United States General Accounting Office
USGS	United States Geological Survey
WRC	Water Resources Council
WRPA	Water Resources Planning Act

References

Ackerman, E. A., 1960, *The Impact of New Techniques on Integrated Multipurpose Water Development*, Committee Print no. 31, US Senate Select Committee on National Water Resources, USGPO, Washington, DC.

Ackerman, E. A. and Loef, G., 1959, *Technology in American Water Development*, Johns Hopkins Press, Baltimore, MD.

Canter, B. D. E. and Cristie, D. R., 1984, 'Water laws and policies' in E. A. Fernald and D. J. Patton (eds.), *Water Resources Atlas of Florida*, Florida State University, Tallahassee, FL, pp. 130–8.

Clark, E. C. II, 1985, 'Trends in water management' in *Water Management in Transition, 1985*, a special report by the Freshwater Foundation, Navarre, MN, pp. 12–14.

Derthick, M., 1974, *Between National and State*, Brookings Institute, Washington, DC.

Federal Emergency Management Agency, 1986, *A Unified National Program for Floodplain Management* (FEMA 100/March 1986), FEMA, Washington, DC.

Foxworthy, B. L. and Moody, D. W., 1985, 'National perspectives on surface-water resources', in *National Water Summary, 1985*, USGS Water-Supply Paper 2300, USGPO, Washington, DC, pp. 51–68.

Gregg, F., 1989, 'Irrelevance and innovation in water policy: lessons from the WRPA' in S. M. Born (ed.), *Redefining National Water Policy: New Roles and Directions*, American Water Resources Association (Special Publication no. 89-1), Bethesda, MD, pp. 11–18.

Hart, G. W., 1971, *Institutions for Water Planning*, National Technical Information Service (PB 204244), Washington, DC.

Hines, W. N., 1971, *Public Regulation of Water Quality in the United States*, National Technical Information Service (PB 208309), Washington, DC.

Muckleston, K. W., 1976, 'The evolution of approaches to flood damage reduction', *Journal of Soil and Water Conservation*, **31** (2), pp. 53–9.

Mumford, L., 1961, *The City in History*, Harcourt Brace & World, Inc., New York.

National Water Commission, 1973, *Water Policies for the Future*, USGPO, Washington, DC.

Outdoor Recreation Review Commission, 1962, *Federal Agencies and Outdoor Recreation*, ORRC Study Report no. 13, USGPO, Washington, DC.

Palmer, T., 1986, *Endangered Rivers and the Conservation Movement*, University of California Press, Berkeley, CA.

Piper, A. M., 1965, *Has the United States Enough Water?*, USGS Water-Supply Paper 1797, USGPO, Washington, DC.

Platt, R. H., 1986, 'Floods and man: a geographer's agenda' in R. W. Kates and I. Burton (eds.), *Geography, Resources, and Environment*, Vol. II, University of Chicago Press, Chicago, IL, pp. 28–68.

Price, T. P., 1971, *Hydroelectric Power Policy*, National Technical Information Service (PB 204052), Washington, DC.

Schilling, K. *et al.*, 1987, *Water Resources*, report on the nation's public works, National Council on Public Works Improvement, Washington, DC.

Sewell, D. R. W., 1976, 'The changing context of water resources planning: the next twenty years', *Natural Resources Journal*, **16** (4), pp. 791–805.

Teclaff, L. A., 1967, *The River Basin in History and Law*, Martinus Nijhoff, The Hague, The Netherlands.

US General Accounting Office, 1988, *Flood Insurance—Statistics on the National Flood Insurance Program* (GAO/RCED 88–155FS), USGAO, Washington, DC.

US Geological Survey, 1983, *National Water Summary, 1983—Hydrologic Events and Issues*, US Geological Survey Water-Supply Paper 2250, USGPO, Washington, DC.

US Senate, 1960, *Fish and Wildlife and Water Resources*, Committee Print no. 18, US Senate Select Committee on National Water Resources, USGPO, Washington, DC.

US Water Resources Council, 1967, *Alternative Institutional Arrangements for Managing River Basin Operations*, Water Resources Council, Washington, DC.

US Water Resources Council, 1968, *The Nation's Water Resources*, USGPO, Washington, DC.

Viessman, W. and Welty, C., 1985, *Water Management Technology and Institutions*, Harper and Row, New York, NY.

Wall, J. F. and Schilling, K., 1985, 'The Corps of Engineers: planning to meet the financing challenge' in *Water Management in Transition, 1985*, a special report by the Freshwater Foundation, Navarre, MN, pp. 21–7.

Wengert, N., 1980, 'A critical review of the river basin as a focus for resources planning, development, and management' in R. M. North, L. B. Dworsky and D. J. Allee (eds.), *Unified River Basin Management*, American Water Resources Association, Minneapolis, MN, pp. 9–28.

Wescoat, J. L. Jr., 1984, *Integrated Water Development*, The University of Chicago Department of Geography Research Paper no. 210, The University of Chicago, Chicago, IL.

White, G. F., 1957, 'A perspective of river basin development', *Law and Contemporary Problems*, **22** (2), pp. 157–86.

White, G. F., 1969, *Strategies of American Water Management*, The University of Michigan Press, Ann Arbor, MI.

CHAPTER 3

New Zealand water planning and management: evolution or revolution

Neil J. Ericksen

Water management in New Zealand is in the midst of rapid change. Indeed, the institutional arrangements for allocating and using all natural resources have been undergoing change of revolutionary proportions since the Fourth Labour Government assumed power in 1984. An analysis of water planning and management in New Zealand must take account of these wider changes and the reasons for them. As well, a narrow analysis of water management to the present time would quickly be out of date without consideration of current proposals for change. Yet, to understand proposals for the future, a thorough review of past policies and practices in water planning and management is essential.

This chapter opens with a broad overview of the environmental, economic and political evolution of New Zealand, highlighting the implications of these factors for land and water management. It provides a context for the analysis that follows. Three sections trace the evolution of comprehensive-integrated water and land planning and management over the past half century around the topics of catchment planning and management, water use planning and management and local and regional planning and management. These topics emerged after the Second World War as somewhat distinct and parallel developments. The need to build connections between each of them led to a complex legislative and administrative structure that, in spite of good intentions, was only partly successful in facilitating a comprehensive or holistic approach to integrated land and water management (Mitchell, 1987). The final section draws together the strands of the analyses, focusing on the situation that had evolved by the 1980s. It then considers new government proposals for comprehensive-integrated resource management that seemed destined to guide New Zealanders into the twenty-first century.

Water planning and management in context

Physical environment

New Zealand is an elongated country of two main islands comprising 267,000 km^2, 1,500 km south-east of Australia. Tectonic forces have created axial mountains that run the length of the South Island and much of the North Island (see Figure 3.1a). About 70 per cent of the country is above 300 m. Catchments are relatively small. More than half the 76 major catchments of over 500 km^2 are under 1,500 km^2. The largest catchment in the North Island is 14,349 km^2 (Waikato River) and in the South Island 21,960 km^2 (Clutha River).

The country intersects moist prevailing westerlies in temperate latitudes (34°S to 47°S), and receives relatively high annual rainfalls, particularly west of the ranges (see Figure 3.1b). The succession of anticyclones and depressions that traverse the country at regular intervals also brings rains from the north-east and south-east. Intensive short-duration rains often result in soil erosion on even moderate slopes, and flooding on the plains. These conditions have been exacerbated by large-scale clearing of once abundant forest and draining of swamplands (see Figure 3.1c).

Economic development

The main productive activities in the nineteenth century were the extractive industries of gold, coal and timber, and primary industries based on pastoral farming (wool, meat and dairy). Grassland farming continues to dominate overseas earnings, although its 45 per cent of total is about 60 per cent that of the 1950s. Since then, forestry, fishing, horticulture and manufacturing have dramatically increased export earnings as successive governments sought to diversify production. These industries have accelerated water demand and placed stress on water quality. Until recently, the abundant rain has enabled various users of water to obtain water with relatively little difficulty (see Figure 3.1f).

Since becoming a colony of Great Britain through the signing of the Treaty of Waitangi in 1840 with the indigenous Maori, the population has increased from about 200,000 to 3.4 million, of which 400,000 (11.7 per cent) are of Maori descent. The population has become increasingly urbanized, and at the last census (1986) over half lived in 28 cities of 20,000 or more people (see Figure 3.1e). While almost all cities draw on local waters, the main city, Auckland, is seeking extra-regional supplies, as well as waste disposal sites.

Chemical-based agriculture together with increasing urban-based occupance and industry contributed to the rapid deterioration of natural waters and the recognition of the need for improved water management in

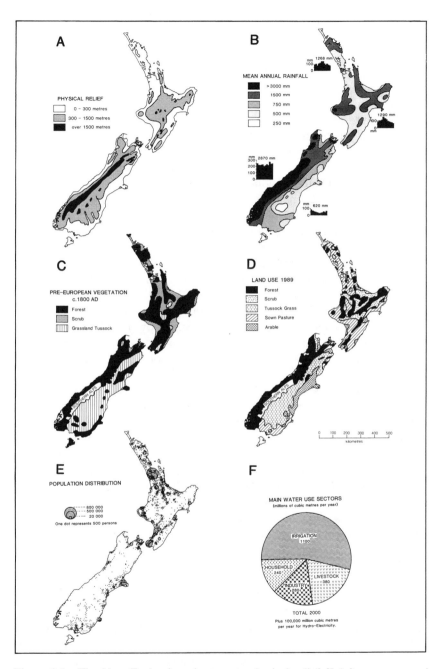

Figure 3.1 The New Zealand environment: physical relief (3.1a); mean annual rainfall (3.1b); pre-European vegetation about AD 1800 (3.1c); land use 1989 (3.1d); population distribution 1986 (3.1e); and main water use sectors 1987 (3.1f)

the 1960s. Similar concerns were held for deteriorating land resources, including loss of forests, rare fauna and scenery.

Political approaches

Enhancement of the 'Welfare State' during the 1930s by the First Labour Government entrenched a mixed economy approach where government intervention was the norm. Although ruled by a 'free-enterprise' National Government for 27 of the 33 years to 1984, intervention policies continued with increasing vigour. Progressive loss of assured markets after Britain joined the European Economic Community (EEC) in the 1960s, and adverse changes in terms of trade into the 1970s, saw government intro-duce a range of economic measures designed to stimulate production and diversification. Measures included protectionism, export incentives and an array of grants and subsidies to stimulate production. The end result was an insulated, very low-growth economy saddled with high inflation and high interest rates. In an effort to stimulate economic growth, wage and price controls were introduced in 1982–4 and government borrowed heavily to finance large-scale natural resource development projects based on local energy sources, including hydroelectric power, gas and coal. Many of the agencies responsible for resource development were also responsible for their management and conservation. Conflicts of interest were rampant, a problem returned to in the final section.

Paradoxically, under the Fourth 'socialist' Labour Government, free enterprise monetarist policies have been vigorously pursued since 1984. A parallel development has been a thorough restructuring of government agencies and their functions. Proposals to restructure regional and local government are in train. Major ministries have been axed, and others created. Many central government departments have been corporatized under the State Owned Enterprises Act 1986. This has had major impli-cations for the management of land and water resources, as nearly 50 per cent of the New Zealand land area is Crown owned. The template for change is still evolving. The likely outcomes are dealt with towards the end of this chapter.

Legislative framework

Water planning and management in New Zealand is complex. Many Acts of Parliament have a bearing on it. Indeed, by 1980, some 70 statutes dealt with environmental matters, and responsibility for administering them involved 20 central government agencies (Barton and Johnston, 1980). The statutes do not devolve responsibility for water resources (or land resources) on any one central agency. Rather, responsibility has been spread through many central agencies, as well as regional and local government, including catchment boards and regional water boards. The major Acts of Parliament relevant to land and water resource planning and management in New Zealand appear in Table 3.1. They are displayed with respect to the powers they confer on government agencies responsible for administering them.

The two statutes most relevant to water and soil management are the Soil Conservation and Rivers Control Act 1941 and the Water and Soil

Table 3.1 Land and water planning and management statutes and powers to act

	Power to:				
Statute	Preserve	Manage	Plan	Finance	Enforce
Clean Air Act 1972		×			
Environment Act 1986	×	×			
Fisheries Act 1908 & amend's	×				
Forestry Act 1949		×			
Fresh Water Fisheries R's 1951					×
Harbours Act 1950		×			×
Health Act 1956				×	×
Land Drainage Act 1908					×
Local Government Act 1974			×	×	
Local Government Amend't 1979		×			
Marine Farming Act 1971		×			
Marine Reserves Act 1971	×				
Marine Pollution Act 1974					×
Mining Act 1971		×			
National Development Act 1979			×		
Queen Elizabeth II National Trust Act 1977	×				
Reserves Act 1977	×		×		×
Soil Conservation and Rivers Control Act 1941	×	×		×	×
Town & Country Plan'g Act 1977			×		×
Water & Soil Conservation Act 1967		×	×	×	×
Wildlife Act 1953	×				
Special Acts: Control of hazardous substances		×			

Source: Collated from Auckland Regional Authority, 1983a

Conservation Act 1967. The 1941 Act provides the basis for the analysis of comprehensive 'catchment planning and management' with respect to soil erosion, flood control and land drainage. Two other Acts are especially important for land and water management and planning at local level. They are the Local Government Act 1974 and the Town and Country Planning Act 1977 (highlighted in Table 3.1). The institutional arrangements they spawned are dealt with in the section on 'territorial regional and local planning'. All resource management law is under review, and the consequences of this for water resource planning and management are the subject of the final section entitled 'towards comprehensive-integrated resource management'.

Catchment planning and management

The rainfall distributions in Figure 3.1b suggest that New Zealand has plentiful rain, well distributed throughout the year, and reliable in occurrence. The main rainfall feature is, however, that it is spatially and temporally variable (Tomlinson, 1976). Temporally, annual variations seem sporadic and quite possibly tied to sunspot cycles. Spatially, annual variations are a function of the north–south orientation of the country intersecting moist prevailing westerlies from the Tasman Sea. The variations are indicated in Figure 3.1a. The need to control and conserve water is evident. This section deals mainly with the problem of too much water, and the need to control it and soil erosion through catchment planning and management.

Towards comprehensive catchment management

The need to control water was evident from the time of early settlement in the second half of the last century, and land drainage and river boards were established. They focused on the more important areas, and ignored the wider catchment. The areal distribution of the boards, which often had differing objectives, was fragmented over limited lengths of a river; in some cases on one side of it.

   Meanwhile, the pace of change on catchment slopes quickened as tussock grass and bush were fired and cleared for farming (Cumberland, 1941). The thin and naturally eroding soils on high country slopes and basins of the South Island were further exposed to erosional forces as their tussock covering was alternately fired then over-grazed by introduced sheep and rabbits. Millions of hectares became significantly eroded, and 50 per cent of soil loss occurred in some areas (Cumberland, 1944a; McCaskill, 1973, p. 49). In the North Island, 5 million ha of bush were cleared between 1870 and 1910, often from slopes underlain by soft mudstones that quickly eroded upon exposure to frequent and intensive rainfalls.

As investments on floodplains increased and were subsequently damaged by flooding, public pressure for new water control works mounted. This in turn led to calls for more comprehensive management. Severe floods and soil erosion in the 1920s and 1930s and a growing realization beyond the experts that land clearance, soil erosion and flooding are linked were the catalysts for New Zealand's first attempts at comprehensive catchment management (Acheson, 1968; Cowie, 1957; McCaskill, 1973; Poole, 1983).

Legitimizing the concept. District committees under the Public Works Department (later the Ministry of Works and Development) were established in 1937 to examine the problem in conjunction with scientific committees. Their work gained urgency as severe floods and erosion ravaged the east coast of the North Island in 1938. In arguing for an integrated approach to land and river control, the committees were supported by advice solicited from scientists in the United States of America.

Originally, it was intended that there be separate Acts for soil conservation and river control (Acheson, 1968, p. 19). For a decade, staff in the Public Works Department had focused on the need for flood control legislation. This probably influenced the Minister of Public Works so that the Bill that went forward to Parliament was restricted to that end. However, heavy lobbying by soil conservators in the Parliamentary select committee considering the Bill, together with a visit by its members to areas suffering severe erosion, resulted in combined legislation: the Soil Conservation and Rivers Control Act 1941 (Howard, 1988; McCaskill, 1973, pp. 22 and 105; Poole, 1983, p. 24). This Act made New Zealand 'one of the first countries in the world to recognise, through legislation, the interrelationship between land and water resources, and the importance of management on a catchment-wide basis' (Burton, 1985, p. 132). It provided for 'unified control both at national and district level' (Acheson, 1968, p. 19). This little modified Act remains, after nearly 50 years, the main tool for comprehensive catchment management.

Administrative structure. The 1941 Act made provision for establishing a centrally co-ordinated regional administrative structure to pursue the objectives of conserving soil resources, preventing damage by erosion and flooding and using land in a manner tending towards the attainment of these objectives. To achieve them, a Soil Conservation and Rivers Control Council (SCRCC) was established at the national level.

The SCRCC was responsible to the Minister of Works, and was serviced by its technical and administrative staff. The SCRCC could call on relevant government agencies for assistance (see Figure 3.2, p. 61). An Interdepartmental Coordinating Committee was established to facilitate the co-ordination of soil conservation work and research.

Responsibility for catchment planning and management was delegated by the SCRCC to new regional catchment boards. The catchment boards (which operated until October 1989) were corporate local bodies. The 1941 Act allowed their membership to vary from between eight and 15, depending on the importance of the district. Membership included elected local residents whose interests were affected by catchment management, and appointed public servants from relevant agencies with skills, such as in engineering, land management, forestry, agriculture and so on (see Figure 3.2). The number of appointed members could not exceed elected members, and was usually in the ratio of 1:2. Two representatives of the catchment boards served on the SCRCC.

Functions of SCRCC and catchment boards. The SCRCC was accorded 14 general functions under Part 1 of the 1941 Act. Distilled, these were to formulate policy, authorize projects and provide funds (from Central Government) for the majority of costs for local erosion control, river control, land drainage, research and data collection and investigations and demonstrations. Part 2 of the 1941 Act gave the SCRRC power to authorize surveys, acquire land in critical areas or to regulate the use of it, carry out demonstration work, and construct soil conservation, river control and drainage works. Most of the powers of the SCRCC were delegated to its regional catchment boards. Construction projects could, however, be accomplished through the Ministry of Works or Department of Agriculture as its agents, particularly where public utilities affected by flooding, erosion and drainage problems were the responsibility of those agencies. Part of project costs could be claimed from individuals benefiting from such works.

Processes and problems of implementation

This bold initiative in catchment planning and management led to dramatic improvements in land use and productivity. Within 30 years of establishing the first catchment board in the Manawatu region in 1943, around 70 river control and land drainage projects had been funded by the SCRCC. By then there were 18 catchment boards covering 80 per cent of the country. They had implemented 30,000 km of channel maintenance, 8,800 km of channel improvements, 3,040 km of stopbank (levee) construction and 5,600 km of river bank protection (Ericksen, 1986, p. 128). A further 55 major projects had been funded by 1988. Together, they 'protected' around 90 flood-prone urban communities. In addition, tens of thousands of hectares of rural floodplains had been drained and protected, thereby dramatically increasing agricultural productivity. On the hills, hundreds of thousands of hectares of erodible land had been treated (McCaskill, 1973; Poole, 1983).

While much was achieved under the 1941 Act, there were, however, many problems with its implementation. The most salient problems are described below under structures, resources and mechanisms for planning and management.

Structures. The administration and management created under the 1941 Act were straightforward enough, but implementation was compromised in several ways. The 1941 Act did not make catchment boards mandatory, and they were rather slow to evolve (Ericksen, 1986, pp. 119–21). Although five had been established in each island by 1945, only four more were added by 1960. They covered only 60 per cent of the country. Around 10 per cent of the country was still not covered by 1988 when an amendment to the 1941 Act made them mandatory. By then there were 20 catchment boards. There were good reasons for not including a mandatory clause in the 1941 Act. Quite simply, it was politically expedient at the national level to leave the decision to adopt catchment board management of an area up to the affected parties, that is, the local people. Thus, catchment boards emerged only in those areas where people perceived they were needed. Key factors against adopting catchment boards included arguments over the basis of representation, reluctance to pay additional rates, difficulties in developing rating classifications and objections to the powers conferred on catchment boards by the 1941 Act.

Another problem of implementation from the Act itself was its narrow mandate. The main objective of the Act was for managing the problems of soil erosion, land drainage and flooding, rather than for water and/or land planning in the wider sense. In addition, while the Act made provision for the integration of soil and water management in the catchment system, integration was confined to these three problems. Thus, most uses of water, as well as problems of water quality, were ignored. This constraint was largely remedied with passage of the Water and Soil Conservation Act 1967, in which nation-wide coverage for the consideration of all water uses was made compulsory by 1973.

In a related sense, demarcation of water control among various administrative authorities also influenced the success of the catchment boards. For example, irrigation remained the responsibility of the Ministry of Agriculture and Fisheries. For a time in the late 1950s, soil erosion also fell under their sway. The control of flood and erosion hazards also became a statutory responsibility of regional and territorial local authorities in legislation beginning with the Town and Country Planning Act of 1953.

Resources. In early years, staff and material resources for operating catchment boards were limited. There were few people in New Zealand specifically trained in catchment work, particularly in soil engineering. The boards therefore relied on other governmental agencies for support, such as the Ministry of Works and Department of Agriculture. As staffing increased, so too did disciplinary or professional demarcation. The division between river engineers and soil conservators was especially clear (Acheson, 1968; McCaskill, 1973).

Howard, who rose to be Director of the Water and Soil Directorate, Ministry of Works and Development, notes that this division between water engineer and soil conservator occurred in both the servicing agencies of the SCRCC and the regional catchment boards. These professionals 'rarely met on common ground' (Howard, 1988, pp. 105–6 and 117–18). He observes that the barriers between river engineers and soil conservators were not broken down until the early 1970s. This was achieved through the process of a major review of progress in catchment management that was instituted by the SCRCC in 1969. The report stemming from the review appeared in 1973, and in it the Chairperson, Mr L. Poole, stated that 'The collaboration that has accompanied this review is perhaps the most valuable feature in regard to the development of policy' (Poole, 1973, cited in Howard, 1988, p. 118).

Critical to the establishment of the regional catchment boards was the power given them (via the SCRCC) to collect rates from people in districts benefiting from catchment management schemes. These rates paid for the administration costs of the catchment boards, as well as the local share of constructing and maintaining various works, typically 25 per cent of total costs. Rates for catchment works were levied on a graduated scale, and lands classified according to the degree of direct and indirect benefit derived from a scheme. The remaining funds for local projects came as a grant from Central Government through the SCRCC (Ericksen, 1986, pp. 122–4).

Mechanisms for catchment planning and management.   Although forward planning for comprehensive catchment control was provided for in the legislation, it developed slowly and was variable in its application. The types of plans, and problems encountered in implementing them, involved region-wide planning, catchment-wide planning and land use capability surveys. The management tools adopted by catchment boards reflected those typically employed in developed countries, including by-laws and land use controls (Acheson, 1968; Poole, 1973).

REGION-WIDE PLANNING

The 1941 Act enabled staff in catchment boards to prepare comprehensive schemes (in effect, plans) covering their districts or regions, but rarely were these produced before the 1970s. Indeed, Section 128 of the Act required that:

Every catchment board as soon as after it has been constituted and not later than such date . . . prepare and submit to the Minister [SCRCC] a general scheme for preventing or minimizing damage within its district by floods and erosion.

Regional planning for a complete catchment control scheme would need to have involved: consideration of general physical features of the catchment; carrying out a land capability survey and determining desirable changes in land use; consideration of the hydrological characteristics of the catchment; a survey of lowlands affected by flooding and drainage problems; preparation of proposals and estimates for remedial works; and, economic evaluation of proposed schemes (Acheson, 1968, p. 51).

These schemes should have formed an important element in total catchment planning. Howard (1988) gives insights into why Section 128 was not enforced. First, in the early years, boards were poorly resourced and information for planning and management was very limited. Professional demarcation and rivalry between soil conservators and water engineers also prevented integration, and therefore hindered a comprehensive approach to water and soil planning and management. Perhaps more importantly, the boards were created to deal with specific major problems of soil erosion and flooding, and were therefore of necessity crisis-oriented and reactive, rather than forward looking and anticipatory. In these circumstances, planning for the total district for which the catchment board had responsibility seemed a luxury.

CATCHMENT-WIDE PLANNING

Comprehensive planning on a catchment-wide basis was also slow to emerge. It was hindered by lack of resources for carrying out data collection and surveys, and by difficulties in simultaneously gaining the support of upstream and downstream beneficiaries of catchment-wide schemes. As Acheson (1968, p. 173) noted, ratepayers in hill catchments were sometimes reluctant to join with ratepayers in the lower valleys where the bulk of expenditure was involved and vice versa. They argued over relative expenditure and benefits, or over classification and rating. In comparison with lowlands, it was often more difficult to classify fully hill country catchments, assess on-site and off-site benefits and determine equitable rates of subsidy. In addition, the control of flood problems was often seen by both catchment board staff and the public as more urgent than the control of soil erosion, reflecting a difference in impacts between the concentrated flood and more diffuse erosion.

The opportunity to resist implementation of catchment control schemes was formally built into procedures of application. The process of public notification of catchment schemes, and the right of individuals to object at a public hearing, were requirements under the 1941 Act. This democratic process could mean long delays in starting a scheme and/or in implementing all its parts. For example, the comprehensive Waihou Valley Scheme was first proposed in the late 1960s, but was only 50 per cent complete when serious regional flooding and erosion struck in 1981 (Ericksen, 1986; Harris, 1980; Smith, 1984).

To encourage comprehensive catchment planning, the SCRCC altered the pattern of grants to catchment boards. The highest grant of 70 per cent of costs was awarded for proposals that emphasized catchment-wide control schemes. Independent river schemes gained 60 per cent, and isolated works only 40 per cent of total costs. Initially, this move was compromised by the professional rivalry noted above. Later, catchment-wide planning was affected by a Treasury Department recommendation to Government in 1971 that projects worth more than $500,000 undergo a benefit cost analysis whereby successful projects had to meet an internal rate of return of at least 10 per cent. This was raised (temporarily) to 15 per cent in 1978. This requirement compromised catchment-wide planning as boards picked the most productive parts of a catchment to ensure some measure of government funds. This meant opting for short-term river control projects rather than longer term catchment treatment and land use management (Ericksen, 1986, pp. 123–8).

LAND USE CAPABILITY SURVEYS

In order to integrate the many measures appropriate for managing various forms of soil erosion and associated water problems, a suitable planning tool was needed. This was found in the Land Use Capability Survey system developed in the United States in the late 1930s (Klingebiel and Montgomery, 1939; Cumberland, 1944b). Various surveys were conducted in the 1950s, but it was not until 1963 that the SCRCC formally adopted an appropriate land classification system, and a further six years before catchment board staff could refer to a *Land Use Capability Survey Handbook* (Ministry of Works, 1969).

The SCRCC did, however, have a policy for providing catchment boards with land inventory and land capability data as a basis for planning catchment conservation as early as 1947, but it had limited personnel to enable delivery (McCaskill, 1973, p. 192). By 1961, this limitation led the SCRCC to provide catchment boards with a subsidy to help their staff gather the necessary information. By the 1970s the SCRCC was meeting all costs to catchment boards of obtaining data related to Land Use Capability Surveys where the minimum size of area was around 20,000 ha and the problem was of national significance. For their part, the catchment boards met the cost of all data gathering related to capability surveys preceding the design of soil conservation works (McCaskill, 1973, p. 192).

LAND USE CONTROL

Particularly important in controlling soil erosion and related water problems on hill country land were farm conservation plans. Policy supporting this management tool was adopted by the SCRCC in 1956. Farm conservation plans were to be employed where multiple, integrated conservation practices were needed to deal with erosion that affected a substantial portion of a farm or run-holding. Subsidies for the scheme were conditional on the farmer adopting a mutually agreed-upon plan of management and finance for the whole property. Conservation needs were established through a land capability survey. Activities were integrated and scheduled over five years, and both costs and results monitored.

Another management tool was provided by a 1959 Amendment to the 1941 Act. Sections 34 and 35 gave very strong powers to catchment boards to control unwise use of land. The power to serve notice on owners misusing their lands was, however, subject to appeal heard by a tribunal. Initially, the approach was to use persuasion through education, rather than appeals to such powers. As abuse of the land continued under both private and public enterprise, the use of Sections 34 and 35 increased. In his retrospective review of catchment management, Howard (1988, p. 106) reports that these two sections 'developed into one of the main planks in our system of catchment management'. While this was so for catchment treatment in rural areas, Sections 34 and 35 seem to have been used sparingly in urban areas against flood hazard. In addition, they were difficult to enforce and monitor.

BY-LAWS

Under Section 149 of the 1941 Act, catchment boards, as local authorities, could pass by-laws relating to the creation and maintenance of protection works, including use of land. In early years, the latter was restricted mainly to land uses in designed floodway zones. Later, in the 1970s, new developments in protected urban areas were increasingly required to have land and/or buildings elevated above the design level of stopbanks. Adoption of this practice was slow because water engineers did not always promote it, and local government resisted being regulated (Ericksen, 1971; 1986).

Water use planning and management

In many ways, the problem of dealing with water use and quality has a similar history to that of soil erosion and flood control, although the evolution from a fragmentary to integrated approach emerged later.

New Zealand has around 30 times more runoff entering the sea than is needed to eliminate deficits that result whenever evapotranspiration exceeds rainfall (Howard, 1985, p. 135). This abundance of water has in part been responsible for a cavalier attitude to its use. Fast flowing, readily available natural water was seen as the best means for diluting and dissipating effluents from rural and urban areas. The danger of pollution was seen as remote.

As competition for clean water among users increased, the need to plan water use more carefully was increasingly realized. An example of both growing public concern, and political recalcitrance is shown in a debate that has raged this decade in Wellington, the capital city, over milliscreening sewage into the Wellington Harbour (Anon., 1985a, pp. 11–12). In Auckland, the largest city, harbour pollution compares with large cities in the United States. Thus, even today, 30 years after the introduction of water pollution control measures, the cavalier attitudes of yesteryear persist amongst significant water users.

Towards integrated water planning and management

In response to various and growing demands on water, a wide range of legislation and statutory responsibilities evolved in a piecemeal fashion from early years to guide state, county, municipal, and private users. Legislation was, however, either sector-oriented or use specific. Acts guiding government departments included reference to water use where appropriate. Other Acts made provision for territorial local authorities to construct and operate water works and sewerage systems. Some large cities had their own Act, as did some major industries.

Legitimizing integration.   By the 1950s there was considerable public debate over the quality of the New Zealand environment. A great deal of concern was expressed in the 1950s by fisher people and recreationalists demanding clean water. Their concerns were picked up by the Marine Department, which was at that time responsible for fisheries. It might well have been the Health Department, because most community water supplies were of relatively poor health standard. A 1965 survey showed 46 per cent of the total population were receiving unsatisfactory water supplies. The outcome over concern for clean water was the Water Pollution Act 1953. Administered by the Marine Department, the Act made provision for restricting water pollution through establishment of a Water Pollution Council. A decade later the Water Pollution Regulations 1964 enabled waters to be classified according to their uses. However, very little had been achieved in solving water allocation and water quality problems by the 1960s, and the professional, public and political debates intensified (Cowie, 1964).

In 1962, the Nature Conservation Council was created to advise the government on matters pertaining to preserving the quality of the natural environment, including water. Conflicts over resource development and conservation intensified throughout the 1960s as improved communication led to a widely informed public willing to challenge the plans of central government bureaucracies.

In this context, the Minister of Works established a committee involving ten government departments to examine the water use problem in 1963. To push things along, the New Zealand Institute of Engineers chose 'water use and control' as its theme for the first of several influential water conferences (Water Conference, 1964; 1970; 1976; 1982; and 1988). It was opened by the Minister of Works, the Hon. P. B. Allen, who stated:

foremost, some agency must take on the job of completely co-ordinating all water administration to ensure effective action . . . There is also the need for allocating functions to someone for the many matters that are not now the responsibility of any particular organisation—water conservation, water allocation, further aspects of water quality and general administration (and) comprehensive research . . . Assembled here are scientists, engineers, water use experts, civil authorities, and others . . . I invite everyone . . .to sink personal parochial differences and assist me in placing sound legislation on the statute books . . . which will be of great assistance to the generations yet to come (Allen, 1964, pp. 3–6).

Many of the contributors to the conference were engineers and scientists who came from government agencies. Howard (1988, p. 107), gives three reasons for holding the conference: problems of water use competition, failure of existing structures to deal effectively with large polluting industries and a 'fear of the unbridled power of the Crown [Central Government] riding roughshod over other potential water users'. Recommendations from the Conference (Water Conference 1964, 1964) supported the Minister, and within 30 months, Parliament had passed the Water and Soil Conservation Act 1967. This Act furthered New Zealand's reputation for producing innovative land and water management legislation.

Administrative structure. The Water and Soil Conservation Act 1967 did not replace the Soil Conservation and Rivers Control Act 1941, but was linked to it through overlapping objectives. The 1967 Act aimed to promote a national policy for natural water by: (1) making better provision for the conservation, allocation, use and quality of natural water; (2) promoting soil conservation and preventing damage by flood and erosion; (3) providing and controlling multiple uses of natural water and drainage of land; and (4) ensuring adequate account was taken of the needs of primary and secondary industry. An important amendment to the 1967 Act in 1981 added recreation, wildlife and scenic values to the list.

Implementation of the 1967 Act was through a two-tiered national and regional administrative structure that extended the one already developed under the 1941 Act. At the national level, the old Soil Conservation and

Rivers Control Council (SCRCC, 1941) and Water Pollution Council (1953), together with a new Water Allocation Council, were brought under the umbrella of a new National Water and Soil Conservation Authority (NWASCA). The authority and three councils together formed the National Water and Soil Conservation Organisation (NWASCO). The Chairperson of the authority (NWASCA) was the Minister of Works. The authority and three councils were serviced by a new Water and Soil Division, one of several divisions established in the Ministry of Works (see Figure 3.2).

Under the 1967 Act, catchment boards were reconstituted as catchment boards and regional water boards (e.g., Hauraki Catchment Board and Regional Water Board). The regional water boards had responsibility for carrying out the task of comprehensive management of natural water in the field. As evident in Figure 3.2, the new catchment cum water boards continued the institutional relationship with the regional offices of the Ministry of Works and Development, established under the 1941 Act.

Various sector and inter-agency groups with interest in water had representation on both NWASCA and the catchment boards cum regional water boards, much as was described in the section on catchment management (see Figure 3.2). (Where the joint 'catchment board and regional water board' entity is being referred to from now on, the shortened title of 'catchment cum water board' will be used.)

Functions of NWASCA and regional water boards. Extensive powers were given to NWASCA to carry out its many functions. These functions are given in Section 14 of the 1967 Act as: (1) to examine problems concerning, and make plans in respect of: the allocation and quality of natural water; the control of erosion on the banks of rivers, the shores of lakes and the seashore; the control of flow and flooding in and from rivers and lakes; conservation of natural water; all enactments and rules of law relating to natural water and the needs of fisheries and wildlife and all other recreational uses of natural water; (2) to advise the Minister from time to time as to what enactments are desirable to ensure the most efficient administration of natural water and the conservation of soil and natural water in the national interest; (3) to co-ordinate all matters relating to natural water so as to ensure that this national asset is available to meet as many demands as possible and is used to the best advantage of both the country and the region in which it exists; (4) to guide national and local administration of natural water and soil conservation and the application of knowledge thereby; (5) to guide and encourage research in matters relating to natural water and the application of knowledge thereby acquired; and (6) to promote the best uses of natural water including multiple uses and to allocate natural water between competing demands.

Most of these powers and functions of NWASCA were delegated to the catchment cum water boards. Unlike the functions of the catchment

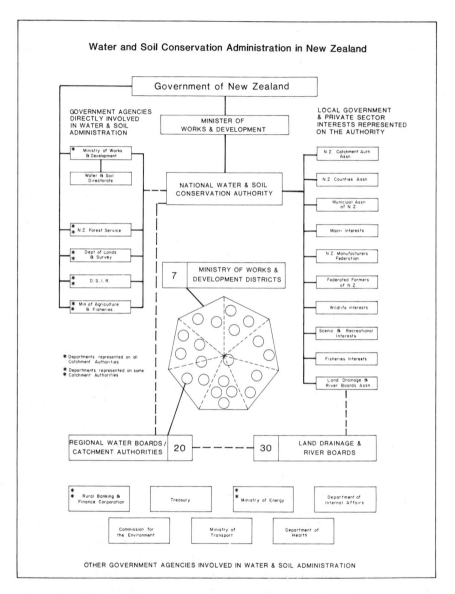

Figure 3.2 Water and soil administration in New Zealand in 1972. The multi-tiered structure integrated central, regional and local sector interests. The NWASCA was created by the 1967 Act, adding two new Councils to the SCRCC created in 1941. In April 1972, the Water Pollution Council and Water Allocation Council were combined into the Water Resources Council (shown in the diagram). In 1984, the councils were disestablished, and in 1988 NWASCA and the Ministry of Works and Development were disestablished leaving the water boards/catchment authorities and land drainage and river boards. (Source: Anon., 1984, p. 20)

boards, those of the new regional water boards were mandatory. All New Zealand had to have coverage by 1973. The distribution of the 20 regional water boards that resulted appears in Figure 3.5a. While the regional agencies had statutory functions and duties, they were required to adhere to the directions of NWASCA and the three councils, and to co-operate with them in developing working policies. Grants from government facilitated this process (Howard, 1988, p. 110).

Processes and problems of implementation

The 1967 Act was a commendable attempt at instituting integrated water management in New Zealand. It had many features aimed at bringing about co-ordinated water management: a good system of government grants to entice co-operation, a democratically elected local system of management and powers to act on a wide range of functions.

While much has been achieved during the more than two decades that the system has operated, it has, however, been fraught with many problems that have impeded efficient implementation. A number of issues were canvassed in government reviews of the system, and piecemeal changes followed. These and other problems eventually stimulated drafting of a new Bill in 1985 combining the 1941 and 1967 Acts, but the Bill has been overtaken by a wider review of resource management law that is still in progress. The various problems can be dealt with under the following headings: structures, resources and mechanisms for planning and management.

Structures. The structural and administrative complexity created by the 1967 Act is apparent from Figure 3.2. At the national level it was top heavy. Consequently, the Water Allocation Council and Water Pollution Council were fused into the Water Resources Council in 1972. Then, in another more wide ranging shake-up in 1984, the organization (i.e., NWASCO) was disbanded when the two remaining councils were dissolved, leaving only the authority (i.e., NWASCA), but with expanded representation. The nature of government and private sector interests also altered in 1984, the latter better reflecting in the administrative structure the long neglected interests of recreation, wildlife, scenery and wilderness values and the interests of the Maori people. The overall process of restructuring could be characterized as reduced central administration, and increased devolution of responsibilities to the regions. While this could be argued as being a good development, it can also be seen as having some limitations, as will be described later (Conway, 1984).

Another problem to emerge was that NWASCO and the catchment cum water boards seemed too beholden to the power of the Ministry of Works and Development. This bounding influence is indicated in Figure 3.2 (Anon., 1984). The Wellington Head Office of the Ministry of Works and Development overlays NWASCO, and the seven district offices of the Ministry overlay the 20 catchment boards and regional water boards. In two cases, boards fell within two or more works districts. The communication of proposals was required to move through the relevant district and head office of the Ministry of Works and Development, to the relevant council and NWASCA. This caused delays in approvals and funding for local works.

In addition, in the 10 per cent of the country where there was no catchment board coverage (under the 1941 Act), but regional water board coverage (under the 1967 Act), administrative functions and processes became confusing as the district offices of the Ministry of Works filled the gap.

Resources. When catchment boards first evolved in the late 1940s and early 1950s, the regional offices of the Ministry of Works were extremely valuable in providing resources to them. As catchment boards expanded, however, this need lessened over the years, and the oversight of the Ministry became increasingly irksome (Ericksen, 1986; Howard, 1988). Added to this irritation was the fact that the Ministry not only controlled the funds for the granting system, but through its representation at all levels was also seen as somewhat omnipotent. Furthermore, as local property owners were reluctant to pay rates in support of catchment boards in some regions, the administrative rates were so low that they had to be subsidized by central government through the Ministry of Works. This exacerbated feelings of dependency and paternalism among people in the catchment cum water boards (Howard, 1988).

Whereas the 1941 Act enabled catchment boards to collect rates or levies to help defray costs of administration and soil, flood and drainage works and maintenance, no such provision was made in the 1967 Act for the water resource planning and management functions of the new regional water boards. This severely hampered implementation of water resource management. The boards needed new staff with a wide range of new skills. Not only were such people scarce, but even when available the boards had financial difficulty employing them (McKenzie, 1976). Not until a 1987 amendment to the 1967 Act could regional water boards impose a water charge on users for the costs of managing water resources (Bates, 1988, p. 95).

Another resource problem was the bias in representation, both of the elected and appointed members of the boards, and the technical officers employed by the boards. Most members and officers were focused on catchment management, and it took some years before changes in representation began to reflect an interest in water resource management.

McKenzie, President of the New Zealand Catchment Association, wrote:

> Engineers now had to consider whether the water retention value of swamps might
> not have more value than the new land that could be reclaimed if they were
> drained. Soil conservators now had to consider not only soil loss, but also the ability
> to slow down run-off and the retention of water . . . Most regional water boards
> appointed a water resources officer as head of the Water Department of the board,
> but there was considerable soul searching about the qualifications needed for this
> appointment . . . If water allocation was the main problem (in the region) then a
> hydrological engineer would be the obvious choice . . . If . . . polluting discharges,
> then a chemist would be the obvious choice (McKenzie, 1976, pp. 90–1).

A decade later, many catchment cum water boards were still not attuned
to the 1967 Act. Becoming attuned was facilitated by the restructuring that
combined the Water Allocation Council and Water Pollution Council into
one Water Resource Council in 1972. This led to better representation
on council, and moves towards better liaison with the regional boards
(McKenzie, 1976, pp. 88 and 90).

Mechanisms for water planning and management.   The 1967 Act provided
three main tools for the regional water boards to carry out their functions:
water classifications, water rights, and prosecutions and penalties. Amendments to the legislation added: (1) by-laws related to groundwater (1973);
(2) water conservation orders on wild and scenic rivers (1981); and (3)
water management charges (1987). Over time, NWASCA added several
non-statutory measures, such as Water Allocation Plans (1973), Baseline
Water Quality Surveys (1974) and Water and Soil Resource Management
Plans (1978). All of these measures presented problems for water planners
and managers trying to bring about an efficient water use system. As
time passed, and experience was gained, efforts to solve these problems
emerged.

WATER CLASSIFICATION

The most specific planning power given to regional water boards was the
water classification system provided by Section 14 of the 1967 Act. (The
power to classify had already been established under the old Water
Pollution Act 1953.) The 1967 Act enabled the Water Pollution Council
(later the Water Resources Council) to set minimum standards of quality
for each class. This would, in effect, provide a national plan for water
quality against which the regional water boards could determine the
permissible quality and quantity of an effluent, and the best location for the
prospective discharge of waste (Cowie, 1976, pp. 52–3).

   To be effective, a system of water quality management based on receiving water standards needs good information on aquatic systems and the
interactions between land use and water systems in order to define standards that will apply to given water, and predict the effect of wastes on the

quality of receiving waters. As already noted, resources for this sort of work were very limited.

Criticisms of the water classification included its being imposed by the Water Pollution Council on regional water boards without drawing on local knowledge, being unmindful of local plans for land use and failing to provide information on its construction, or on how it could be used by the boards (McKenzie, 1976, p. 88; Taylor, 1976, pp. 6–9).

Poor communication from the Water Pollution Council in Wellington to its regional water boards resulted in a classification with which many users disagreed. Providing an immutable standard ignored long-term changes to water quality caused by changing land uses in a catchment and the various means by which wastes could be dispersed in natural waters. Applying the classification proved a complex business. By 1976, the system of classification had been tested in law (Skelton, 1985; Williams, 1980). For example, two cases brought against the Water Resources Council by the Southland Acclimatisation Society (1975) and Southland Skindivers Club (1976) found that the approach of the Council in not adopting the existing water quality standards as an element in decision making was contrary to the Act. Following this, the classifications were therefore found to be erroneous in that they could have the effect of degrading the existing quality of water. These decisions limited further work in this field, and in 1989 many regional water boards still did not have classified waters.

WATER ALLOCATION PLANS

Non-statutory Water Allocation Plans were proposed by NWASCA at the end of the 1960s. They were to depict the way water in a catchment should be allocated amongst defined uses and competing demands. Functionally, they were to provide regional water boards with a framework within which to consider applications for rights for users to take water. Restrictions on water take would relate to the level of supply. The plan would also guide water users on the availability of water in their areas (Cowie, 1976, p. 52). The lack of funds and trained staff in early years impeded production of Water Allocation Plans, but equally important was the dearth of data on which plans could be based. Later, guidelines and financial grants from NWASCA were provided as inducements to regional water boards to implement the plans.

WATER RIGHTS

Complementing the proposal for water classification plans in the 1967 Act was the water right system for managing water use. A key element in the 1967 Act was that all natural water belonged to the Crown. Under Section 21, the control for damming any river, diverting, taking, or using any natural water, or discharging natural water or waste into natural water, was vested in the Crown. A system of water rights was established as the means through which these activities would be controlled. The water right system was the main management tool provided by the 1967 Act, for it was the means by which both water classification and water allocation plans could be implemented. The water classification would provide the planning guidance on the quality standard of receiving waters, and the water rights the management means by which effluents would be controlled in order to meet those standards. Operationally, the 'system proved full of legal pitfalls because there could be many claims on one body of water and many uses of it already in existence' (Poole, 1983, p. 24).

The process of granting, deferring, or declining water rights to applicants was delegated to managers in the regional water boards. The process also involved public notification, water resource investigation, water board decision making and appeal (Bates, 1988, p. 94). Obviously, this could be a lengthy and costly process. The process generated most controversy over applications for water rights to dam rivers and mine minerals (McKenzie, 1976, pp. 93–6; Wilson, 1982).

The main deficiency of the water right system was that it focused on individual applications, whereas to be effective in water management, individual applications needed to be examined in the context of the wider interest. To help overcome this problem, there was a move towards comprehensive water rights and the production of water allocation plans and/or overall resource management plans.

This move towards comprehensive water rights was supported through Section 22 of the 1967 Act which outlined several types of water right, including a 'general authorisation'. In addition, through Section 16, regional water boards could delegate powers to territorial local authorities. The granting of a general authorization or comprehensive water right is an example. This would allow a city council, for example, to approve of and/ or carry out multiple minor uses without the need for specific water rights from the regional water board. Control by the regional water board was maintained through the need for the city council to produce plans, such as for storm-water, and to renew their general authorization every five years. This system meant that the territorial local authority could deal directly with land developers on water-related matters.

PROSECUTION AND PENALTIES

Where unauthorized activities affecting natural water, or non-compliance with a water right occurred, the regional water board had authority under Section 34 of the 1967 Act to pursue prosecution. The maximum fine was NZ$2,000 (US$1,300 at 1989 exchange rate). For years, there was an exceeding reluctance on the part of regional water board managers to prosecute. There were at least three reasons for this. In earlier years, it was felt that a long period of education was needed to lead people into behaviour more consistent with the water quality objectives of the 1967 Act. However, in contrast to the information campaigns in the 1950s aimed at high country farmers causing soil erosion, little effort was actually made by the Water Resources Council of NWASCA to educate the public on water use and quality. Rather, it was felt that over time, changes in infrastructure would generally lead to the opportunity for implementing improved water uses and discharges. Second, it was considered unfair to prosecute a factory for a large polluting spill, while in the same region hundreds of dairy farmers, for example, washed cowshed waste into stream systems each day. (Since 1984, dairy farmers have been required to spray wastes onto pastures.) Third, bringing a prosecution could be costly, and even where it succeeded, the fine under the 1967 Act was so low as to offer no deterrent at all to the offender. After a long period of abuse, the maximum fine was eventually raised in 1987 to NZ$150,000 (US$100,000).

Section 154 of the 1941 Act also allowed for prosecution in relation to soil and water matters. However, as with several other Acts seeking to improve environmental quality, the fine was minuscule ($1,000).

CONSERVATION ORDERS

It is evident from previous discussion that making operational the water classification and water right systems proved difficult. In effect, the 1967 Act provided a framework for decisions on water allocation through water rights based on the interlocking concepts of multiple use, beneficial use and balanced use. As Guthrie and Parker (1988, p. 211) explain, multiple use implies new water rights being considered in view of existing uses, no use should forestall another use and adverse impacts from water use are expected. Beneficial use implies a productive or useful purpose. Balanced use implies weighing the benefits of a new use against the adverse effects on water flow. Operationally, these concepts threw up a number of problems that were tested in tribunal hearings and in the courts.

For example, Section 20(6) of the 1967 Act states that due regard should be given for recreational needs, safeguarding wildlife, natural features and scenery, and avoiding severe impacts on streams and rivers. In practice, however, the prevailing view in granting water rights was that if the use was beneficial and productive, then negative impacts on conservation values, while regrettable, were acceptable (Guthrie and Parker, 1988, p. 212). Also, while essentially single-purpose dams could be built on a river, conservationists could not gain a water right to 'manage' a river as an in-stream fishery.

These sorts of judicial decisions, together with highly controversial government hydroelectric power projects, such as on the Clutha River, caused conservationists to lobby for an amendment to the 1967 Act so that primacy could be given for nationally important conservation values, such as wild and scenic rivers (Guthrie and Parker, 1988; Powell, 1978; Wilson, 1982).

The National Government relented in election year, 1981, and any public authority specifically constituted under any Parliamentary Act could make application for a national water conservation order, as could any Minister of the Crown (Water and Soil Conservation Amendment Act 1981). The process requires public notice to be given by NWASCA, written submissions called and organized public hearings held. A committee of inquiry appointed by NWASCA investigates and reports. Any draft order is made public and submissions received are then considered by a Planning Tribunal. A report and recommendation passes to the Minister. The Minister is not bound by NWASCA or Tribunal decisions. Any party can appeal against a decision of the Minister in the High Court on questions of law, and then to the Court of Appeal. Upon the recommendation of the Minister, the Governor General then makes the national water conservation order. The full process is therefore long and costly.

The first conservation order was placed on the Motu River in the wilderness of the Bay of Plenty in 1983. The Motu had been prospected since the 1950s for possible dam building sites. The next conservation order involved part of the Rakaia River in Canterbury. It proved much more difficult to place since primacy for conservation values would impede extensive use for irrigation and other productive purposes on the plains. The decision went to the Court of Appeal (Bates, 1988, p. 214; Burton, 1985, pp. 129–31).

WATER CHARGES

The 1967 Act did not include any reference to the financing of regional water board functions. The catchment cum water boards were, therefore, forced to rely on the rates they gathered under the 1941 Act for purposes of flood, soil and drainage control, and grants from central government through NWASCA. Not until 1987 did this situation change when an amendment to the 1967 Act allowed catchment cum water boards to charge users for the cost of managing water resources. As Bates (1988, p. 95) observed, since the charge can be based on water quantity, it has the capacity to raise the awareness of right holders of the value of their water rights and, therefore, should prove a useful conservation measure.

This water charge does not have to stretch to the provision of water supplies and water treatment for communities because supply and treatment are the responsibility of territorial local authorities, not catchment cum water boards.

WATER AND SOIL RESOURCE MANAGEMENT PLANS (WASRMPS)

The review of NWASCA, between 1969 and 1972, identified the need to integrate land and water planning, including water use and quality (Poole,

1983). To meet the multiple objectives of the 1941 and 1967 Acts, catchment cum water boards were being required to produce several types of plans on a catchment-wide or sub-catchment basis. This highlighted the need for integrating land and water matters in a wider geographical framework.

By the mid 1970s, integrated planning was being discussed under a bewildering raft of titles: Regional Management Plan (Cowie, 1976, p. 51; Gibson, 1976, p. 122), Regional Water Plan, Regional Water and Soil Management Plan, and Water and Soil Resource Management Plan (Howard, 1988, p. 118). It was not until the late 1970s that the concept was clarified. By then, the Water and Soil Directorate of the Ministry of Works and Development was 'vigorously promoting' the use of Water and Soil Resource Management Plans (WASRMPS), and subsidies were being given by NWASCA to stimulate their preparation by the catchment cum water boards. Not only would such plans serve the needs of the boards, but they would provide an important framework for advising territorial regional and local authorities on water and land matters.

In practice, WASRMPS tended to be done only where large-scale development projects created an immediate need. Thus, by 1985, nearly 20 years after the Water and Soil Conservation Act 1967 had been passed into law, only eight of the 20 regional water boards had them (Howard, 1985, p. 136). A major reason for the ambivalence to this sort of integrated planning was the continued myopia of the physical scientists and engineers in the catchment cum water boards. Resource planning was given low priority, as more specific and immediate problems were focused on. Even in the 1980s, it was common to find only one or two planners appointed to a board of, say, 60 staff. There are even instances of an engineer taking on the role of planner and being made senior in status to the professional planner on the board (Ericksen, 1986).

A key problem was that these sorts of plans were not required in law, and while attempts were made to include them in draft legislation in the 1970s and 1980s, nothing came of it (Anon., 1985b; Howard, 1988, p. 118). Strangely, although the need for integrated WASRMPS seems to have been widely accepted, and even vigorously promoted by the Water and Soil Directorate, NWASCA did not produce a policy on the matter until late 1987, just before it was disbanded by government. This led some boards to develop their own alternatives (Bowden and Priest, 1988, p. 33).

Territorial authorities: planning and management

Of fundamental importance to problems of implementing water planning and management in New Zealand is the role played by territorial regional and local authorities. They have responsibility for land use and many water-related functions, which in turn relate to functions of the catchment cum water boards. The territorial local authorities are particularly im-

portant because they have had a significant influence politically on the way
in which land and water planning and management have evolved.

Territorial local government

Territorial local government made up of counties and municipalities (hereafter called local authorities) was established in New Zealand in 1876 at the
same time that Provincial Government was abolished (Municipalities Act
1876 and Counties Act 1876). Local authorities were given power to collect
a land tax from property owners to carry out their statutory functions
(Rating Act 1876). By 1988, local government had grown to include 80
counties and 131 municipalities. Their statutory responsibilities had
not changed greatly, including managing water supply, sewage disposal,
rubbish disposal, roads, parks and reserves, land subdivision, land use
planning and community development, along with health and building
inspection, civil defence, pensioner housing and libraries (Local Government Act 1974).

As early as the 1920s, planning was made explicit for municipalities in
the Town Planning Act 1926, and was strengthened and extended to
include the counties in the Town and Country Planning Act 1953. This Act
required local authorities to prepare district planning schemes to guide
growth and development in forward periods of 20 years with five-yearly
reviews. The scheme was to have as its general purpose the wise use and
management of the resources of the district, and the direction of the
control of its development, in such a way as would most effectively
promote and safeguard the economic, cultural, social and general welfare
of the people and the preservation of amenities of the area (Section 18). A
1973 amendment required matters of national importance to be considered
in the schemes, particularly those relating to natural resources and the
environment.

A code of ordinance was required for the administration and implementation of the district scheme, and maps were to be included to illustrate
proposals for development of an area. The schemes had to deal with
matters outlined in the Second Schedule of the Act. Most items had direct
or indirect relevance for water and land planning, such as: (1) zoning of
areas for specified purposes, (2) designating reserves for afforestation and
catchment purposes and open space, (3) control of subdivision, including
for flooding and (4) water supply, sewage and waste disposal.

Territorial regional government

Recognition of the need for local planning to be done in a wider regional
context was evident in the many attempts by both major political parties

to create territorial regional authorities (hereafter called regional authorities).

The planning Acts of 1926 and 1953 both made provision for regional planning. In the Town and Country Planning Act 1953 these provisions were: (1) to carry out surveys of natural resources, and of present and potential uses and values of all lands in the region in relation to national, regional and local development, public works, services and amenities and then (2) to prepare a Regional Planning Scheme. In this scheme, regional authorities were to deal with matters outlined in the First Schedule to the Act, these being: (1) communications and transport, (2) land use, (3) public utilities and (4) amenities. The last three were particularly important, as the detailed matters in the Schedule related to water, soil and land.

Processes and problems of implementation

The problems of implementation of planning and management of land and water resources through the statutory procedures of territorial government were legion, and included difficulties in generating planning schemes and cross linkages between the territorial authorities and the catchment cum water boards.

Planning schemes. The local authorities moved with glacial speed into district scheme planning. By 1970, only 70 per cent of them had a prepared scheme. Even then, many simply adopted the model code produced by the Town and Country Planning Division of the Ministry of Works rather than developing one sensitive to local needs (Ericksen, 1986, p. 148).

The creation of regional authorities to carry out regional planning was dependent upon the agreement of local authorities within the agreed upon region. Local authorities have always been very parochial and protective of their interests, and therefore vigorously resisted government moves to develop regional authorities (Roberts, 1976). Thus, few were formed until the new Local Government Act 1974 made them mandatory by 1979.

The Labour Government of 1972–5 made provision for strong Regional Councils with power to collect rates, appoint staff and have elected members. Attuned to the disaffection of local politicians and vested interest groups, the succeeding National Government revised the legislation so that a much weaker form of regional government could be chosen: the United Council. United councils were entirely dependent on the local authorities which appointed their members to the united council, provided finance to it through a levy on themselves and seconded their staff to it. United councils were, therefore, cheap as well as dependent. Unlike regional councils, united councils could not charge rates, appoint their own staff, nor have directly elected public members.

Not surprisingly, by 1989 there were only three Regional Councils: Auckland (1963), Wellington (1980) and Northland (1987), whereas before 1977 there were six. Of the 22 councils in existence by 1988, 19 were united councils. An illustration of the weakness and dependency of the latter is seen in the case of the Waikato United Council. While it serves one of the richest agricultural and energy resource areas in the country, with the fastest growing city, Hamilton, at its centre, it had only two part-time staff seconded to it in seven years of operation to 1985. There has been little improvement since.

Cut-and-fill development leading to subsidence and landslippage, suicide subdivisions in flood-prone areas, river and harbour pollution and little regard for aesthetic values, were some of the problems that emerged in local authorities throughout New Zealand as growth and development accelerated into the 1970s. Furthermore, little headway had been made with respect to regional planning. Consequently, the planning and management Acts were strengthened. Under the Local Government Act 1974, nation-wide coverage of regional authorities became compulsory. Under the new Town and Country Planning Act 1977, the regional planning scheme was to 'prevail' over the district scheme plan. Commendable though this was, it had little practical value in regions where local authorities opted for united councils.

To cope with the growing problem of harbour degradation, the 1977 Act also included a new Maritime Planning Schedule. Matters to be dealt with included preservation and conservation of harbour resources, maintenance of water quality, avoidance or reduction of damage by flooding, erosion and silting and the relationship between marine activities and the activities on adjacent land. In like spirit, the Town and Country Planning Act 1977 was more specific than its 1953 predecessor on matters pertaining to district and regional schemes, in terms of the relationship between land and water uses. It drew attention to the need to avoid or reduce danger or damage caused by flooding, erosion, landslip and subsidence. It was also clear on the need to preserve and conserve land with wildlife, scenic and recreational values.

Linkages with catchment cum water boards. Regional catchment cum water boards contain many local authorities. There are also portions of one or more regional authority within their boundaries. Many of the functions of regional and local authorities were also matters of concern to catchment cum water boards, although the means for dealing with specific issues could differ in emphasis. One specific example is land use planning of local authorities versus river control of catchment boards in the case of flood problems. While there was opportunity for working hand-in-glove on common problems, like flooding, operationally this failed for a variety of reasons. To overcome this, the legislation relevant to territorial authorities on water matters was progressively strengthened.

As a case in point, Section 4(3) of the Town and Country Planning Act 1977 linked the purposes and objectives of regional and district scheme planning to that of soil and water management. More specifically, the 1977 Act states: 'In the preparation, implementation, and administration of regional, district, and maritime planning schemes . . . regard shall be had to the principles of the *Soil Conservation and Rivers Control Act 1941* and the *Water and Soil Conservation Act 1967*'. In addition, a representative from the catchment cum water board was to serve on the mandatory committee of the regional planning committee. Amendments in 1978 and 1979 to the Local Government Act 1974 required local authorities not to approve any scheme plan where it was satisfied that the land was not suitable for subdivision because of flooding and other natural hazards, or where subdivision would aggravate such hazards. Section 274 of the Local Government Amendment Act 1978 linked it to the Town and Country Planning Act 1977 with respect to subdivision development, district scheme plans and hazard avoidance planning. The Local Government Amendment Act 1979 required local authorities to refuse building permits in hazardous areas, such as from flooding and landslip, unless protective measures were taken (Section 641(2)).

There were, however, countervailing forces in the local authorities against these sorts of legislative constraints. For example, local authorities lobbied to have weakened Section 641 of the Local Government Amendment Act 1979 through a 1981 amendment. But there were other forces as well. Some of them stemmed from the statutory provisions themselves. To illustrate these forces, the case of urban flood-hazard reduction will be briefly explained.

Urban flood hazard. In New Zealand, an integrated approach for reducing flood hazard has long been advocated (Ericksen, 1971; 1975; 1982; 1986). A raft of statutes supports measures that modify the event and its effects through catchment treatment and flood control schemes, modify loss-susceptibility through land use management and community preparedness and modify losses through spreading the burden in time and space using insurance and various forms of relief and rehabilitation (Ericksen, 1986). Yet in spite of the legislation, land use management through the local authorities is weakly adopted, while the other measures are strongly adopted. This outcome is the consequence of many complex factors. One of the most important is a structural problem created by the legislation itself.

Under the Soil Conservation and Rivers Control Act 1941, catchment boards were created to employ specialists who could modify the flood event and its effects through catchment control, and they were given the finances to do it. A similar observation can be made for the institutional structures and measures dealing with flood losses. But local authorities, while having statutory authority for dealing with flood hazard under the

Town and Country Planning Act 1953/1977 and amended Local Government Act 1974, had many other competing problems to take their interest
and resources. More important, central government subsidies were not
directed to local authorities as an inducement for reducing flood-loss
potential through land use management. Subsidized options are invariably
more attractive than non-subsidized options.

Another important reason for the lack of progress made by local
authorities in reducing flood hazard through land use management was
attitudinal (Ericksen, 1971; 1975; 1986). It was the belief amongst many
local politicians and other community influentials that land use management would limit development and therefore rateable income for use in
stimulating further growth. It was this factor that caused advice from
catchment boards to local authorities on floodplain planning to be compromised, even spurned. Common amongst local planners and politicians
has been the view that flood problems are for catchment boards to resolve
through the provision of flood protection works. Few seemed to appreciate
that flood control works escalate disaster potential by encouraging increased development in the protected and adjacent unprotected areas.
However, many decision-makers did seem ignorant of the nature of flood
hazard (as opposed to flood events) and of the measures that local
authorities can take to reduce it, in spite of their statutory role in preventive planning.

In this atmosphere, it is little wonder that only 30 per cent of the 103
flood-prone municipalities in New Zealand surveyed in 1982 had mapped
flood risk (Ericksen, 1986, pp. 226–31). Most of these maps contained very
limited information and were not very accessible to the public. Most
respondents from the municipalities excused their poor performance as
being the result of a lack of resources to do the mapping and/or being taken
to court should areas once shown in the scheme plan as safe be then shown
to be at risk.

On the other hand, catchment boards held the expertise on matters of
flooding, and it is an indictment on their performance that they did not
produce flood-hazard information maps of direct relevance for land use
planning in local authorities within their districts. As well, many in the
catchment boards saw the resolution of flooding as building more river
control works. This attitude complemented that of people in the local
authorities who saw flood problems as belonging to catchment boards.

Since the early 1980s, members of the National Water and Soil Conservation Authority (NWASCA) and staff in its servicing agency, the
Water and Soil Directorate, Ministry of Works and Development, came to
see that an integrated or unified approach was needed for reducing the
nation's growing flood-loss protection cost bill. In late 1987, NWASCA
implemented a floodplain management policy that required catchment
boards to work closely with local authorities in producing a multi-method
approach to local flood problems. As an incentive, grants would be made

available to local authorities for land use management and related techniques. Any subsidies for flood alleviation measures would be conditional upon measures being consistent with a Floodplain Management Plan. These plans were to evolve from studies that assessed the flood hazard, identified the design flood, evaluated the full range of adjustments for responding to flood hazard and provided guidelines for development within the floodplain (Koutsos, 1988; NWASCA, 1987). The Floodplain Management Plan would fit into the overall Comprehensive Catchment Management Plan which would identify all resources in a catchment, indicate progressive uses over time and assess the impacts these would have on water flows and quality.

NWASCA was disbanded on 1 April 1988, four months after it adopted the unified floodplain management policy. At that time, it was midway through revising all its water and soil policies. The advisory and granting functions of NWASCA have been taken over by a new Ministry for the Environment. Other functions have been disbursed to other agencies or are suspended.

Towards comprehensive-integrated resource management

This final section summarizes progress in comprehensive and integrated water and land planning and management in New Zealand, then outlines the main lines of current political reforms that seem destined to influence the structures, functions, administration and processes of water planning and management in the decades ahead.

From innovation to enervation?

The Soil Conservation and Rivers Control Act 1941 set the context for innovative catchment planning and management. Despite efforts to broaden the concept to include comprehensive water and soil planning, catchment boards focused mostly on physical works as they attracted the highest grants. Enjoining water use and quality under the Water and Soil Conservation Act 1967 made some progress towards improving the data base for water resource planning, but integrated land and water management on a catchment-wide and/or regional basis never totally emerged. A rather narrow vision of their catchment cum water board mandate and the parochialism of territorial local authorities severely compromised the inherent opportunities for comprehensive-integrated management presented in these two Acts.

Efforts to pressurize local government into organizing for regional planning through the Planning Acts of territorial government were similarly unspectacular, as were efforts through legislation to require territorial authorities to plan for integrated water and land management. Attempts to link land and water planning and management responsibilities of territorial authorities to the catchment cum water boards through amendments to the Town and Country Planning Act 1977 and Local Government Act 1974 were helpful, yet superficial, and did nothing to streamline what had become a complicated system of rules and institutions.

The sheer complexity of the sytem is illustrated in Figure 3.3. The chart displays a method for producing a comprehensive catchment plan that places water and land matters in a wider ecological and cultural context. The plan, called a Land and Water Management Plan (LAWMAP), stemmed from detailed investigations of the catchment surrounding Upper Waitemata Harbour in the early 1980s. It was done by the Auckland Catchment Board and Regional Water Board, which forms part of the structure of the Auckland Regional Authority (Auckland Regional Authority, 1983b). The plan was prescriptive, not only for the catchment in question, but also for demonstrating to local authorities within the region, and other interested agencies, its use for comprehensive-integrated water and land planning and management. Procedurally, the LAWMAP would be used to promote its policies in the regional planning scheme, and to guide day-to-day decision making in respect of water rights and routine catchment activities of the catchment cum water board and territorial local authorities. It is pertinent to note that the Auckland Catchment Board and Regional Water Board is one of the most progressive in the country.

In addition to problems of integrating local and regional approaches to water and land planning, were the activities of central government departments and ministries. These activities did not always set an example in wise resource use, and frequently cut across the operations of the catchment cum water boards.

In these circumstances, land was drained and important wildlife habitats become threatened. Hundreds of kilometres of embankments and channels were enlarged against floods, yet flood losses grew. Subsidized land clearance and farming flourished in uplands, but soil erosion continued. Catchments were developed, but effluents from rural and urban land uses flowed into water systems. The quality of many water bodies was improved, but progress was limited.

In less than 50 years, water and land planning and management had moved from a state of innovation to enervation. It was time for a change. Change had been foreshadowed in reviews of soil and water resource management structures and legislation in the 1970s and 1980s. A new Water and Soil Bill was drafted in 1985 to consolidate existing legislation and to facilitate evolution of the organization and techniques of water management. To enhance integration, it proposed a new statutory planning process at two levels: the preparation of Water and Soil Policy Statements as part of mandatory regional planning schemes, and Water and Soil Management Plans to deal with details of management in the

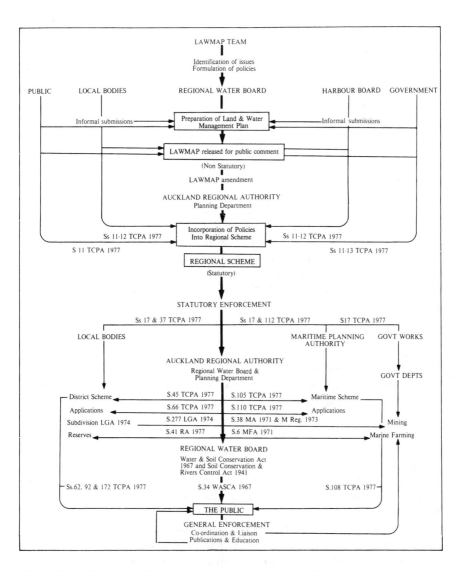

Figure 3.3 Methods of implementation of the Land and Water Management Plan (LAWMAP) for the Upper Waitemata Harbour Catchment (Auckland). This was the most advanced blueprint for integrating land and water planning and management in New Zealand. The process is made difficult by the multiplicity of relevant agencies and Acts (e.g., S11 TCPA 1977). (Source: Auckland Regional Authority, 1983b, p. 12)

regions. There would be greater devolution of functions from NWASCA to catchment cum water boards. The thinking in the Draft was reflected in policies being revised by NWASCA at its time of termination. In terms of the legislation, the intent of the Bill has been overwhelmed by reforms that have completely restructured the institutions of central, regional and local government, and the legislation underpinning all natural resource planning and management. In practical terms, a number of its prescriptions have been pursued to varying degrees by the catchment cum water boards, and can be expected to carry over into the new Regional Councils that replace them in October 1989.

From evolution to revolution?

In his opening address to the Water Conference 1988, Dr Roger Blakeley, Secretary of the new Ministry for the Environment, quoted an eminent political historian, Professor John Roberts, as saying that, since the Fourth Labour Government assumed power in 1984, 'New Zealanders have seen the greatest changes anywhere, anytime in any democracy' (Blakeley, 1988, p. 3). He went on to state that 'we are undoubtedly in the midst of a revolution in policy reform of the legal and institutional framework for resource management in this country' (Blakeley, 1988, p. 3).

Central administration. Underpinning the change was a new ideology: an open and more competitive market and, therefore, the stripping of protectionism and subsidies from the existing welfare state system. Where appropriate, central government agencies would become cost effective, profit-oriented, State Owned Enterprises, as a prelude to privatization. In both private and public sectors, structural economic reforms have wrought succeeding waves of newly unemployed.

Under the old administrative system, central government agencies were separated by resource (water, forest, energy, land), each with multiple functions (policy, regulatory, commercial and conservation). It was easy for an agency to compromise one function in favour of another, for example, commercial interests over conservation interests in managing the forests of a given area. In a related way, one agency would generate a problem while another set about resolving it, as when the lands department would clear forests for farming causing erosion, while a catchment board would plant trees to stop it. Both operations enjoyed generous subsidies.

Under the new system, agencies are separated by function and have responsibility for integrated resource management. The contrasting systems are illustrated in Figures 3.4a and 3.4b. It is argued that the latter will give greater clarity of objectives and accountability for performance than the former (Blakeley, 1988, p. 5). For example, policy advice to the government on the environment now comes from the new Ministry for the

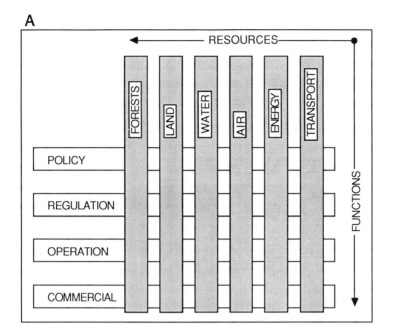

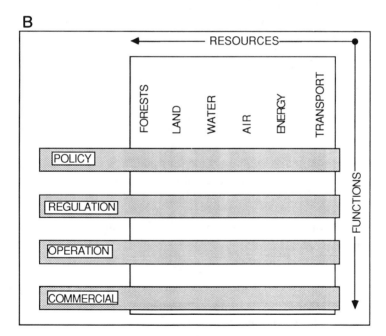

Figure 3.4 Radical change in resource management in New Zealand, 1986. Previous administrative structure had each agency separated by resource with integrated functions (3.4a); new administrative structure has separation by function with integrated resource management (3.4b). (Source: Blakeley, 1988, p. 6)

Environment. It is to give integrated advice on land, air, water, minerals and other resources. The Ministry's role is to see that management of natural and physical resources is integrated with ecological, economic, social, cultural and spiritual values. But the Ministry for the Environment does not have operational, regulatory or commercial functions. In the environmental domain, operational functions are being performed by the new Department of Conservation (Environment Act 1986). It is the advocate for environmental and conservation values, and is responsible for managing the nation's heritage areas, such as wetlands, forests, wildlife and historic places. Commercial functions for natural resource development are being performed by government agencies, such as the Land Corporation and Forestry Corporation, and the private sector. Regulatory and operational functions for natural resource management are to be carried out at regional and local levels through territorial governments.

Regional and local administration.   Reform of regional and local government is no less profound. The seventh Local Government Commission (since 1947) was given power by government to push through reform of the functions, structure, funding, organization and accountability of the 22 territorial regional governments, 213 territorial local governments and over 450 special purpose local authorities, 20 of which are catchment cum water boards (Local Government Amendment Act 1988). Policy has been established and the reorganization is to take place in the local government elections of October 1989. There will be two main types of territorial authorities: (1) directly elected regional councils with a major role in resource management functions and (2) directly elected district councils carrying out functions at the local level.

At the regional level, the existing dual structure of catchment authorities/ regional water boards (see Figure 3.5a) and regional councils/united councils (see Figure 3.5b) will merge into new regional councils whose geographical boundaries will coincide with major water catchments (see Figure 3.5c). There will be 14 regional councils: nine in the North Island and five in the South Island. In addition to considering communities of interest in determining boundaries, the Commission has tried to achieve commonality of jurisdictional boundaries for major natural resource management strategies, land and water planning, maritime planning and other environmental planning matters (Local Government Amendment Act 1988).

The core regulatory functions of the new regional councils will include those of the old territorial regional authorities: regional planning, civil defence and, where appropriate, maritime and urban transport planning, plus the water and soil functions of the catchment cum water boards. Other regulatory functions will probably be added once all of the resource law has been reviewed and consolidated. There also are service functions that will apply to the new regional councils. These are permitted under the Local

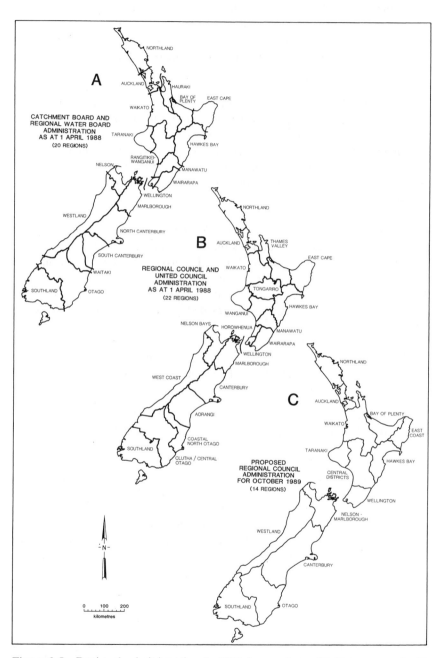

Figure 3.5 Regional administration in 1988 for: water and soil resources in 20 catchment cum water boards (3.5a); natural and physical resources in 22 regional and united councils (3.5b); and proposed combined functions of 3.5a and 3.5b in 14 regional and resource management councils based on major catchments (3.5c)

Government Act 1974, and include regional water supply, regional waste
and sewage disposal and regional reserves. These services transcend local
boundaries (Smith, 1988). Many of the special purpose local authority
functions will be absorbed by regional councils and/or shared with new
district councils (e.g., pest destruction, land drainage and electricity reticu-
lation, but not health and education).

The district councils will emerge from amalgamations of existing territorial local authorities into fewer, larger units. Functions will not greatly
alter, except every attempt will be made to avoid conflicts of interest
among planning, resource management, service delivery, trading and other
functions. Separate committee structures will be established to achieve
this. Trade-offs among objectives are to be made in an explicit and
transparent manner.

Resource Management Law Reform (RMLR). To help meet the challenge of structural change, all laws relevant to natural resources are being
reviewed by staff in the Ministry for the Environment. The time-frame for
the review, with new law as the end-point, is May 1987 to April 1990. The
RMLR review is being done in conjunction with the review to reform local
government (LGR). Both reviews are being controlled through the same
Ministerial committee.

The Government has already agreed to certain principles being included
in the third and final phase of the RMLR review (Ministry for the
Environment, 1988). With respect to water and soil, some of the most
important principles include:

1. Providing for greater integration of consent processes, with resource
   management consents going through territorial government in the first
   instance;
2. Requiring regional government to have Regional Resource Management Policy Statements which would: provide for integrated land,
   water, air and pollution management at regional level; and be adhered
   to in administering lower level plans;
3. Enabling the preparation of Water and Soil Management Plans;
4. Phasing out existing use privileges and mining privileges which have
   created so many of the problems in water management in the past.

Implications for future water management

It is difficult to project what will happen in the future when the present is in
such a state of fundamental change; some would say confusion. But clues
can be gleaned both from current intentions and past experiences.

The analyses of past attempts at creating innovative water and land management structures show three common problems inhibiting comprehensive and integrated water and soil planning and management in New Zealand: professional biases of staff in responsible agencies, institutional demarcation between agencies with supposedly shared interests and lack of finances to implement appropriate methods of analysis and management.

Restructuring involves rationalization, and implies a net reduction in staffing levels. New career opportunities will open; others will close. There will be strong competition for occupational places in the new system. People with attitudes attuned to the traditional role of manager in either catchment/water board or united council/regional council and local authorities will seek entry to the new system, along with those who are attuned to the newly defined needs and goals. The extent to which people with traditional attitudes will occupy influential niches and thereby impede progress towards the integrated management that is being designed remains a moot point. Structures are important, but it is people who count.

It is difficult to predict whether institutional demarcation will be any less problematic under the new administrative structures than under the old. A cause for optimism is that the old structures and laws are being reformed with this demarcation problem in mind.

Lack of financial support for previous functions, and the nature and emphasis of government subsidies, did much to undermine efforts to integrate water and land planning in the past. A 1988 amendment to the Water and Soil Conservation Act 1967 not only allows water users to be charged for water management, but also for rating levels to be increased. This, together with the prevailing user-pays, cost-effective approach of government, should facilitate the multiple resource management functions of the new administrative bodies. That is, savings from restructuring should be directed into improving resource planning and management. It will be interesting to see whether a country of only 3 million people can afford the cost of well developed regional government in addition to central and local government.

New Zealand has a reputation for throwing hastily prepared legislation at its problems then making corrective amendments as experience highlights the shortcomings. The water and land planning Acts of the past provide evidence of this. While the intentions for resource administration and law reform seem commendable, they may be compromised by the breakneck speed with which the reforms are being pushed through.

For any system to work, there must be good quality research and data gathering, and there must be good quality information and education programmes. The demise of NWASCA and the Water and Soil Directorate has seen some of the research functions go to the Department of Scientific and Industrial Research. The functions of information, education and the synthesis of policy and methods seem to have been lost, or at best diffused to the regions. The Ministry for the Environment has picked up policy and funding matters, but seems to lack the resources and capacity to fill the void left at national level. There is now no water and land planning at national level.

Devolution of functions away from central government to the localities where people have to live with their decisions is a fundamental principle of the Fourth Labour Government's approach. While government failure economically and environmentally was common enough under the old system, it will be interesting to see whether market forces and the politics of local government under the new system will enable New Zealanders to fare any better. Let us hope they are not throwing out the baby with the bath water.

References

Acheson, A. R., 1968, *River Control and Drainage in New Zealand*, Government Printer (for Ministry of Works), Wellington.

Allen, P. B., 1964, 'Opening address' in *Water Conference 1964*, pp. 3–6.

Anon., 1984, 'Water and soil conservation administration', *Soil and Water*, **20** (3), pp. 19–24.

Anon., 1985a, 'Wellington city—affluence or effluent?', *Soil and Water*, **21** (1), pp. 11–12.

Anon., 1985b, 'New Water and Soil Bill (1985)', *Soil and Water*, **21** (3), pp. 40–3.

Auckland Regional Authority, 1983a, *Guideline: Legal Aspects of Land and Water Planning*, Auckland Regional Authority, Auckland.

Auckland Regional Authority, 1983b, *Guideline: Comprehensive Catchment Planning*, Auckland Regional Authority, Auckland.

Barton, C. J. and Johnston, J., 1980, *Statutory Functions and Responsibilities of New Zealand Public Service Departments 1979*, Government Printer (for State Services Commission), Wellington.

Bates, B., 1988, 'Water management—state of the art review' in *Water Conference 1988* (Proceedings, Part I), pp. 93–100.

Blakeley, R., 1988, 'Water in society: policy and practice' in *Water Conference 1988* (Proceedings, Part I), pp. 3–28.

Bowden, M. J. and Priest, R., 1988, 'Water planning and management' in *Water Conference 1988* (Proceedings, Part I), pp. 29–40.

Burton, J. R., 1985, 'Water resources development: for what? for where? for why?', *New Zealand Agricultural Science*, **19** (4), pp. 128–134 (Dan Watkins Lecture).

Conway, M., 1984, 'Gone but not forgotten', *Soil and Water*, **20** (3), pp. 35–39.

Cowie, C. A., 1957, *Floods in New Zealand: 1920–1953*, Government Printer (for Soil Conservation and Rivers Control Council), Wellington.

Cowie, C. A., 1964, 'Water-quality management' in *Water Conference 1964*, pp. 29–38.

Cowie, C. A., 1976, 'Essentials of sound water management in New Zealand' in *Water Conference 1976* (Proceedings, Part I), pp. 47–57.

Cumberland, K. B., 1941, 'A century's change: natural to cultural vegetation in New Zealand', *Geographical Review*, **31** (4), pp. 529–54.

Cumberland, K. B., 1944a, *Soil Erosion in New Zealand*, Government Printer (for Soil Conservation and Rivers Control Council), Wellington.

Cumberland, K. B., 1944b, 'Survey and classification of land', *Trans. NZ Royal Society*, **74**.

Ericksen, N. J., 1971, 'Human adjustment to floods', *New Zealand Geographer*, **27** (2), pp. 105–29.

Ericksen, N. J., 1975, 'Is a comprehensive flood plain strategy needed in New Zealand?', *Soil and Water*, **12** (2), pp. 17–18, 33–4.

Ericksen, N. J., 1982, 'Shaping flood-loss reduction policies in New Zealand', *Geography and Development Policy*, New Zealand Geographical Society (Conference Proceedings), Wellington.

Ericksen, N. J., 1986, *Creating Flood Disasters? New Zealand's Need for a New Approach to Reducing Flood Hazard*, NWASCA Misc. Publication no. 77, Wellington. (Summarized in *Soil and Water*, **22** (1), 1986, Special Issue: Flood Hazards in New Zealand.)

Gibson, A. W., 1976, 'The role of the National Water and Soil Conservation Organisation in the support of regional planning for water' in *Water Conference 1976* (Proceedings, Part I), pp. 116–22.

Guthrie, J. K. and Parker, D. W., 1988, 'Water and Soil Conservation Amendment Act 1981' in *Water Conference 1988* (Proceedings, Part I), pp. 211–17.

Harris, R. W., 1980, *Waihou Valley Scheme and the Proposed Classified Rating Area*, Hauraki Catchment Board and Regional Water Board, Te Aroha.

Howard, R. K., 1985, 'New Zealand's water resources', *New Zealand Agricultural Science*, **19** (4), pp. 135–8.

Howard, R. K., 1988, 'A New Zealand perspective on integrated catchment management', *Working Paper for Integrated Catchment Management*, Australian Water Resources Council, University of Melbourne, Melbourne, pp. 101–31.

Klingebiel, A. A. and Montgomery, P. H., 1939, *Land Capability Classification*, US Department of Agriculture Handbook, no. 210.

Koutsos, P., 1988, 'Water hazard management' in *Water Conference 1988* (Proceedings, Part I), pp. 137–50.

Local Government Commission, 1988, *Draft Reorganisation Schemes: Reform of Local Government in New Zealand*, Local Government Commission, Wellington.

McCaskill, L., 1973, *Hold This Land: a History of Soil Conservation in New Zealand*, Reed, Wellington.

McKenzie, R. D., 1976, 'The role of regional water authorities (in water resource management)' in *Water Conference 1976* (Proceedings, Part I), pp. 86–99.

Ministry for the Environment, 1988, *People, Environment, and Decision Making: the Government's Proposals for Resource Management Law Reform*, Ministry for the Environment, Wellington.

Ministry of Works, 1969, *Land Use Capability Survey Handbook*, Government Printer, Wellington.

Mitchell, B., 1987, *A Comprehensive-Integrated Approach for Water and Land Management*, University of New England, Centre for Water Policy Research, Occasional Paper no. 1, Armidale, NSW.

National Water and Soil Conservation Authority, 1987, *Floodplain Management Policy*, NWASCA Circular no. 1987/1, Wellington.

New Zealand Statutory Acts and Regulations:

1908: *Land Drainage Act* (1908 no. 96).
1926: *Town Planning Act* (1926 no. 52)
1941: *Soil Conservation and Rivers Control Act* (1941, no. 12)
1953: *Town and Country Planning Act* (1953 no. 91)
1953: *Water Pollution Act* (1953)
1954: *Municipal Corporations Act* (1954 no. 76)
1956: *Counties Act* (1956 no. 64)
1959: *Soil Conservation and Rivers Control Amendment* (1959 no. 48)
1964: *Water Pollution Regulations* (1964)
1967: *Water and Soil Conservation Act* (1967 no. 135)
1974: *Local Government Act* (1974 no. 66)
1977: *Town and Country Planning Act* (1977 no. 121)
1978: *Local Government Amendment Act* (1978 no. 43)
1979: *Local Government Amendment Act* (1979 no. 59)
1981: *Water and Soil Conservation Amendment Act* (1981 no. 123)
1985: *Water and Soil Bill* (Draft)
1986: *State Owned Enterprise Act* (1986 no. 124)
1986: *Environment Act* (1986 no. 127)
1987: *Water and Soil Conservation Amendment Act* (1987 no. 203)
1988: *Local Government Amendment Act* (1988 no. 3)
1988: *Water and Soil Conservation Amendment Act* (1988 no. 47)

Poole, A. L., 1983, *Catchment Control in New Zealand*, Water and Soil Division, Ministry of Works and Development (for National Water and Soil Conservation Organisation), Miscellaneous Publication no. 48, Wellington.

Powell, P., 1978, *Who Killed the Clutha?*, John McIndoe Press, Dunedin.

Roberts, J. L., 1976, 'A historical summary of the implicit national goals in planning and managing water use' in *Water Conference 1976* (Proceedings, Part I), pp. 3–13.

Skelton, P. R., 1985, 'Sharing water—a legal perspective', *New Zealand Agricultural Science*, pp. 145–51.

Smith, D. H., 1984, *Waihou Valley Scheme Five Yearly Review*, Hauraki Catchment Board, Te Aroha.

Smith, D., 1988, 'Institutional reform' in *Water Conference 1988* (Proceedings, Part II), 1.6, pp. 1–11.

Taylor, M. E. U., 1976, 'Water resources management' in *Water Conference 1976* (Proceedings, Part II), pp. 3–13.

Tomlinson, A. I., 1976, 'Climate' in I. Wards (ed.), *Atlas of New Zealand*, 1976, Government Printer, Wellington, pp. 82–9.

Wards, I., 1976, *Atlas of New Zealand*, Government Printer, Wellington.

Water Conference 1964, 1964, *The Use and Control of Water*, Technical Publications Ltd. (for New Zealand Institute of Engineers, Water Resources Committee), Wellington.

Water Conference 1970, 1970, *New Zealand Water Conference 1970* (Proceedings in 3 parts: I Background statements; II Presented papers; and III Conference discussion and conclusions). Second National Water Conference, 12–14 May 1970, Lincoln College Press (for Organising Committee, New Zealand Institution of Engineers and Royal Society of New Zealand), Christchurch.

Water Conference 1976, 1976, *New Zealand Water Conference* (Proceedings in 2 parts: I Presented papers; and II Supplementary papers). Third National Water

Conference, 23–26 August 1976, Christchurch, Organising Committee for the New Zealand Institution of Engineers and Royal Society of New Zealand, with the support of the National Water and Soil Conservation Organisation, place of publication not given.

Water Conference 1982, 1982, *New Zealand Water Conference* (Proceedings). Fourth National Water Conference, 24–26 August 1982, Auckland, Organising Committee for the New Zealand Institution of Engineers and Royal Society of New Zealand, Auckland.

Water Conference 1988, 1988, *Water in Society: Policy and Practice* (Proceedings in 2 parts: Part I Presented papers; and Part II Supplementary papers). Fifth National Water Conference, 15–19 August 1988, Dunedin, University of Otago (for the Organising Committee of the Institution of Professional Engineers New Zealand and Royal Society of New Zealand), Dunedin.

White, G. F. (with Brinkmann, W. A. R., Cochrane, H. C. and Ericksen, N. J.), 1975, *Flood Hazard in the United States: a Research Assessment*, University of Colorado, Institute of Behavioral Science, Program on Technology, Environment and Man, Research Monograph 6, Boulder, Colorado.

Williams, D. A. R., 1980, 'The water resource: some current problems of law and administration', *New Zealand Law Journal*, pp. 499–507 (D2).

Wilson, R., 1982, *From Manapouri to Aramoana: the Battle for New Zealand's Environment*, Earth Works Press, Waiwere, Auckland.

CHAPTER 4

Integrated water management strategies in Canada

D. A. Shrubsole

Introduction

Water has played an important role in the history of Canada and will continue to contribute to the economic and social development of the country (Burton, 1972; Muller, 1985). The management of this and associated land-based resources involves the consideration of many natural and human variables and is becoming increasingly complex. One response to this challenge has been the development of a variety of institutional arrangements to provide for the utilization and allocation of water resources. The provision of water services in Canada has evolved from single purpose, single means strategies based upon a limited range of technical criteria, to multiple purpose, multiple means strategies based upon a wider set of technical, economic, social and environmental criteria (Dorcey, 1987a; Mitchell, 1986a; Quinn, 1977, 1986; Tate, 1981). The management of water resources has become a testing ground for attempts to co-ordinate various levels of government and user groups through diverse institutional arrangements in order to sustain a high quality of life for Canadians.

In spite of the evolution of more integrated strategies in Canada, the 1980s have seen mounting apprehension over the state of water resources and the ability of private and public managers to administer them effectively. Pearse (1986, p. 9), the Chairman of an Inquiry into Federal Water Policy completed in 1985, concluded that the public perceived water policy

as piecemeal, lacking coherence, and hence inadequate to ensure that water will be managed appropriately in the face of conflicting demands . . . As water has emerged as a mainstream issue, so has the general perception that water policy and administration is disorderly, fragmented and weak.

These concerns were echoed by the Science Council of Canada (June 1988) and Environment Canada (1988). Given these comments, it is appropriate to focus upon the institutional arrangements for water management in Canada.

The physical and human context

At first glance, Canadians would appear to have little cause for concern over the management of water resources. The needs of a population of only 25 million people are served by a mean annual discharge of some 105,000 cubic metres per second, which represents nearly nine per cent of global runoff (Pearse et al., September 1985, p. 28). These characteristics of low population and high supply have led some researchers to classify Canada as a water-rich country (WRI, 1986, as cited by McDonald and Kay, 1988). Although Canada is known for its abundant water resources, this is, in fact, a myth (Foster and Sewell, 1981).

The application of averages is meaningless since supply and demand are not evenly distributed over time and space. Over sixty per cent of the runoff drains northwards while ninety per cent of the population and 'the vast majority of the country's industry is concentrated within 300 kilometers of its southern border' (Foster and Sewell, 1981, p. 8). Mean annual precipitation ranges from over 2,000 millimetres on the west coast of British Columbia to less than 500 millimetres in Saskatchewan. A lack of rainfall during the summer of 1988 caused significant financial losses for the grain farmers of the prairie provinces of Alberta, Saskatchewan and Manitoba. In fact, five of the six water-deficient basins are located in the more arid regions of Canada and are dominated by agricultural land uses (see Figure 4.1). According to Foster and Sewell (1981, p. 17), these six regions will face increasing difficulties in the future, and

the difficulties in southern Ontario [which is the most heavily industrialized and populated province] will be especially severe since the demands will be three times the annual reliable flow, and they will constitute the largest set of water demands in the country. The need for vigorous action to avert the imminent crises in these regions is clear.

Despite the reality of present water shortages, the myth of superabundance has meant that attempts to promote more efficient water use have been difficult (Thompson, 1987).

Scarcity is not the only dimension of water quantity problems in Canada. Potential flood problems exist for as many as 200 communities (Environment Canada, 1983). Significant flood damages have occurred in Vancouver (1948), Winnipeg (1950), Toronto (1954) and Fredericton (1974). Flooding has created major problems in several communities throughout the country as indicated by the $70 million in damages experienced by the Yukon territory and six provinces in 1974 (Environment Canada, 1975a).

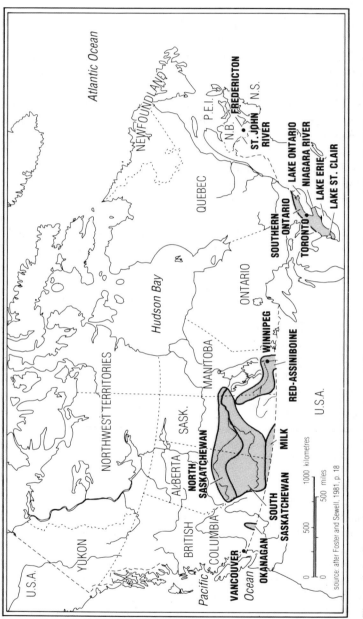

Figure 4.1   Potential water-deficient regions of Canada (Source: after Foster and Sewell, 1981, p. 18)

In addition to water quantity issues, there are significant problems related to water quality. A report published by Environment Canada, which synthesized data from government, academic and scientific sources, commented that

> although this country has not experienced the extreme levels of [fresh] water degradation found in parts of the industrial world, nor the kind of competition for water found in densely populated Third World countries, an overall assessment of the state of Canada's aquatic ecosystems does not suggest grounds for complacency (Bird and Rapport, May 1986, p. 77).

Some of the specific problems are the presence of toxic chemicals in the Niagara River and Lake Ontario, the high contribution of nutrients from diffuse sources to Lakes St. Clair and Erie, the waters of the prairie provinces and the St. John–St. Croix basin in New Brunswick (see Figure 4.1). The construction of dams, interbasin transfers, the loss of critical wetlands and the contamination of groundwater aquifers also posed significant challenges to managers (Cherry, 1987; Rosenberg et al., 1987; Whillans, 1987).

Canada has an extensive shoreline which comprises three oceans and the Great Lakes (see Figure 4.1). These coastal resources are under significant stresses related to offshore oil and gas developments which are found off the east coast of Newfoundland and in the Arctic regions, and onshore activities such as forestry and agriculture (see Figure 4.1) (Harrison and Parkes, 1983). With the exception of Prince Edward Island, every province and territory of Canada shares its water resources with another province, territory, or country (Barton, 1986). This sharing, the uneven distribution of water resources and population, and the competing demands placed upon the water and related land resources make management complex and requires the involvement of all relevant participants.

At a basic constitutional level, Canada has followed the British tradition of governance and adopted a system of responsible Cabinet–Parliamentary government (Doern and Phidd, 1983). Federalism is a central concept in the political system of Canada and is based upon the belief that there is a need to balance national goals and objectives with regional concerns and differences (Saunders, 1986). This fragmentation has served to highlight an 'enduring theme in Canadian federalism: the need for federal–provincial compromise in overcoming these inherent problems' (Romanow, 1986, p. 1).

One central feature of the Constitution of Canada which pertains to the management of water resources is the division of responsibilities between the federal and provincial governments. The ten provinces of Canada have substantial control over water resources within their jurisdiction because the Constitution has conferred the following responsibilities to the provinces: (1) ownership of most of the water resources; (2) ownership over land, mineral and forest resources which impact water resources;

(3) responsibility for property and civil rights; and (4) control over regional and municipal governments beneath them (Cairns, 1987; Pearse, 1986; Rueggeberg and Thompson, 1985; Saunders, 1987). These have enabled all provinces to enact legislation on matters pertaining to water supply and quality, irrigation, drainage, recreation and power generation.

While the Provinces derive their major powers by exercising proprietary and legislative rights, water management responsibilities are shared with the Federal Government which has exclusive legislative jurisdiction over navigation, inland and ocean fisheries, interprovincial works, trade and commerce and international relations. In addition, the Federal Government could intervene unilaterally into any area of provincial jurisdiction if it is deemed to be in the national interest, under powers related to the principle of peace, order and good government (POGG). According to Saunders (1987, p. 468),

> while POGG may support certain interventions by the federal government on specific issues, it would not justify wholesale intervention by the federal level in water management issues otherwise reserved to the provinces.

Thus, the Federal Government has been reluctant to utilize the powers under POGG and infringe in areas of provincial jurisdiction. While the responsibility for water management is shared in the ten provinces, the Federal Government is granted sole jurisdiction to federal lands and water in the Yukon and Northwest Territories, national parks and Indian Reserves.

Another important aspect of Canadian water management is the fragmentation of responsibility which exists within federal and provincial governments. Specific statutes have been passed which provide a legal base to guide water resource development. These levels of government have also created public agencies, usually departments of the environment or natural resources, which consolidate several aspects of renewable resource management (Fisheries and Environment Canada, 1976).

The division of responsibility between and within governments has often led researchers to suggest that effective water management in Canada 'is almost by definition a joint federal–provincial approach' (Quinn, 1977, p. 221).

> The basic constitutional framework for water management in Canada . . . is characterized by a shared, but by no means precise, division of responsibility . . . Jurisdictional ambiguities are not new to Canadian federalism of course, and the typical reaction of the Canadian federal system to such problems is to avoid direct confrontation. This means avoidance of judicial resolution of the problems and the negotiation of compromises in federal–provincial—or occasionally, interprovincial—agreements (Saunders, 1987, p. 471).

Agreements and accords, co-operative standard setting, interministerial co-ordination, advisory bodies and regulatory intervention are mechanisms utilized for federal–provincial co-operation (Lucas, 1986). In addition to these federal–provincial mechanisms, the participation of relevant agencies within a level of government is also required if an integrated water management programme is to be realized.

Saunders (1987) maintained that the legal aspects of interjurisdictional water management have never been as prominent in Canada as in the United States because Canadians have traditionally negotiated agreements rather than relied upon the courts. From a legal viewpoint, the Inter-provincial Cooperatives case provides the 'strongest and most relevant statement on interjurisdictional environmental regulation and manage-ment' (Lucas, 1986, p. 44). The Province of Manitoba attempted to compensate fishermen in the province who suffered decreased harvests as a result of pollution from chori-alkali plants in Ontario and Saskatchewan. In a split decision, the Supreme Court of Canada ruled the Manitoba statute invalid since it attempted to modify civil rights appropriately granted by the legislatures of Ontario and Saskatchewan. The majority of the Supreme Court judges ruled that only the Federal Government could enact this form of legislation (Estrin and Swaigen, 1978). The split nature of this decision indicates that the case has 'little precedential authority' for water management (Barton, 1986, p. 236). Thus, the legal aspects of water management in Canada are not defined clearly and there are few court judgements to guide management activities.

The inclusion of the Charter of Rights and Freedoms in the repatriated Constitution of 1982, could lead to the more frequent utilization of the courts to resolve disputes. Section 7 of the Charter could enable private citizens to call for a judicial review of the management practices and decision-making processes of private and public water managers (Lucas, 1986). Thus, the traditional practice of water management in Canada could shift from one rooted in negotiation and compromise among public agencies, to one characterized increasingly by litigation and confrontation among private and public users.

The treatment of water management activities in this chapter is, of necessity, selective. However, the examples reviewed are illustrative of the approaches and issues associated with the management of this resource. Although not mutually exclusive, the general approaches to the manage-ment of water are: (1) the creation of integrated water management agencies; (2) the implementation of water policies and related legislation; (3) the use of environmental impact assessment and normative planning efforts.

Integrated water management agencies: Conservation Authorities in Ontario

The Province of Ontario provided the earliest example of integrated water management in Canada. The need to employ Second World War veterans and a concern for the degraded state of renewable natural resources prompted the Provincial Government to pass the Conservation Authorities Act in 1946 which provided for the establishment of regionally-oriented, watershed-based agencies. The Act identified their purpose as 'to establish and undertake . . . a program designed to further the conservation, restoration, development and management of natural resources other than gas, oil, coal and minerals'. This broad mandate was one element of the integrated strategy which was to guide the development of Conservation Authorities. Other elements of this strategy included a watershed juris-diction, local initiative, regional resource planning, a provincial–municipal partnership and collaboration with other line agencies (Shrubsole, 1989).

There are 38 Conservation Authorities, most of which are located in southern Ontario (see Figure 4.2). Each is a corporate body, called a Conservation Authority, and consists of individuals who are appointed by member municipalities or the Provincial Government. Approval of projects is required from the Conservation Authority and the Province of Ontario in order to qualify for provincial grants administered through the Ministry of Natural Resources. Member municipalities contribute to the Conservation Authority through funds raised through property taxes. Most Conservation Authorities elect an executive committee which oversees operational aspects of the programmes. Each Conservation Authority is responsible for hiring its own technical and support staff.

While the legislative mandate indicates that Conservation Authorities could undertake a comprehensive programme, their functions have been dominated by the planning and implementation of flood and erosion control works. Collectively, the 38 Conservation Authorities have spent over $1 billion (Canadian) on structural flood and erosion works, have mapped over 21,000 kilometres of flood-prone areas and have main-tained an extensive flood warning system. Other secondary functions include the operation of over 380 recreation areas, the ownership of over 130,000 hectares of land for multiple purposes, the support of soil conser-vation programmes and the sponsorship of conservation education and information programmes (Powell, 1983; Ontario Ministry of Natural Resources, December 1987). In theory, Conservation Authorities were to be comprehensive water managers but have practiced an integrated water management strategy (Shrubsole, 1989).

In 1987, a provincially sponsored Review Committee of the Conser-vation Authority programme acknowledged that, on the whole, Conser-vation Authorities had been effective resource managers, but concluded that significant changes were needed if better management were to be realized (Ontario Ministry of Natural Resources, December 1987). The recommendations of the review committee focused on the structure, and administrative processes and mechanisms of the programme.

The committee was concerned with the ability of some Conservation Authorities to support adequately the existing programmes. The variability among the 38 Conservation Authorities to manage programmes is indi-

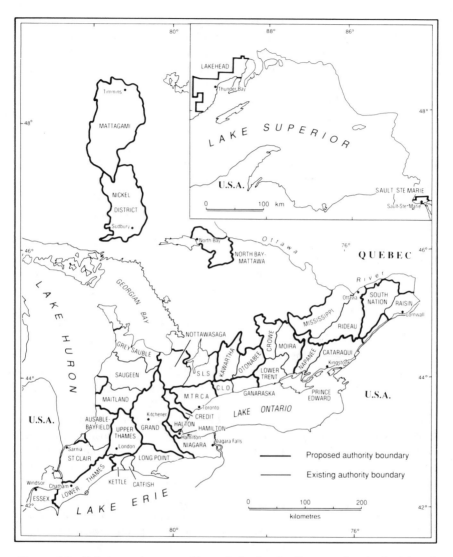

Figure 4.2   Existing and proposed boundaries for the Conservation Authorities of Ontario (1987)

cated by the data below:

Population: range from 9,282 to 2,696,194
Assessments: range from $191 billion to $89,749 billion
Member Municipalities: range from 1 to 42
General Membership: range from 7 to 53
Permanent Staff: range from 4 to 193
Area of Jurisdiction: range from 215 km^2 to 10,933 km^2

In addition, the programmes offered by each Conservation Authority reflect the needs and financial resources of local municipalities and grants available through the Province of Ontario. In order to provide for an adequate distribution of technical and financial resources, it was suggested that the number of Conservation Authorities in southern Ontario be reduced from 33 to 18 (see Figure 4.2).

A second structural issue pertained to the membership of the Conservation Authorities. It was perceived that the large number of representatives on some Conservation Authorities had allowed most of the important decision making to be done by the executive committee. A reduction in membership would eliminate the need for executive committees by most Conservation Authorities. Greater economies of scale and enhanced accountability could be realized through the implementation of these structual recommendations.

There were three processes and mechanisms which were central to each Conservation Authority in addressing integrated water management These were (1) financial arrangements with the Province; (2) integrated resource planning; and (3) collaboration with line provincial agencies.

One central aspect of the provincial–municipal partnership was seen in the development of financial arrangements. Initially, these were based upon a 50/50 split between a Conservation Authority and the Province. Historically, these arrangements have favoured the implementation of structural flood adjustments and often precluded the wider use of other solutions (Shrubsole, 1989). Supplementary grants were established after 1967 in order 'to provide a higher level of assistance to the more rural [Conservation Authorities] where the availability of local financial resources was a limiting factor' (Ontario Ministry of Natural Resources, December 1987, p. 61). The Review Committee maintained that the present system of supplementary grants, which provides a grant of 85 per cent on some projects to 13 Conservation Authorities, is an unbalanced and inadequate form of partnership. It was suggested that regular grants of 40 per cent, 50 per cent or 70 per cent be allocated to each Conservation Authority based upon the assessment and local population. This was believed to be a more responsible allocative mechanism.

A second key aspect of the programme was integrated resource planning. This was achieved initially through the publication of Conservation Reports which were to provide 'a working plan for years to come' (Richardson, 1974, p. 27). The Conservation Reports, which were published when a Conservation Authority was formed, addressed water management problems, related land-based issues, the regional history and economy and the status of current management efforts (Shrubsole, 1989). Unfortunately, these documents were not updated until all Conservation Authorities were requested by the Minister of Natural Resources to submit a 'watershed plan'.

While the completion of watershed plans by all Conservation Auth-

orities in 1983 reaffirmed their commitment to integrated water management, there were a number of weaknesses. First, some watershed plans
were based upon an inadequate amount of research and lacked the strong
support of the Conservation Authority. Second, the future role of Conservation Authorities was identified and justified on the basis of existing
institutional and financial arrangements, yet the adequacy of these delivery
mechanisms was not examined. Third, the commitment of the provincial
government is questionable since a summer student rather than a senior
full-time employee of the Ministry of Natural Resources was responsible
for reviewing the adequacy of the plans. Fourth, a project mentality
surrounds these documents since the Ministry of Natural Resources and
most Conservation Authorities have not attempted to update them on a
systematic and continuous basis. The submission of the watershed plans
became an end in itself, rather than a means to an end (Shrubsole, 1989).

Collaboration with other relevant resource agencies was to be a third
important element of the Conservation Authority programme. This was
important since the responsibility for water management is distributed
among a number of provincial and municipal/regional agencies. For
instance, the Ministry of Natural Resources is responsible for wetlands, the
Ministry of the Environment for water quality, the Ministry of Agriculture
and Food for agricultural drainage, Ontario Hydro for electric power
generation and regional and municipal governments for water supply and
urban development.

The review committee concluded that the broad legislative mandate had
allowed Conservation Authorities to become involved in a wide range of
activities, and had

resulted in overlap with other programs of various ministries (i.e., water quality
and private land extension). The resultant duplication of effort creates public
confusion and some very real inefficiencies in the delivery of these services
(Ontario Ministry of Natural Resources, December 1987, p. 79).

It was recommended that Conservation Authorities should focus their
management efforts on flood and erosion control, low flow augmentation,
the acquisition of wetlands, water quality sampling and the operation of
regionally significant parks. The ability of Conservation Authorities to
respond to a variety of resource issues has been an attribute of the current
programme. Their focus would be restricted, and jurisdictional overlap
with other agencies would be reduced dramatically under this proposal.
Although the implementation of this recommendation could clarify jurisdictional issues, there will continue to be legitimate differences of opinion
regarding the 'best use' of water resources. There has been a strong
tendency by the Provincial Government to minimize and avoid conflict
among public agencies and dodge the need for the development of more
effective conflict resolution or decision-making mechanisms (Shrubsole,
1989).

The integration of water and related land resources has been difficult for Conservation Authorities to achieve due, in part, to the absence of effective planning and conflict resolution mechanisms. For instance, the Conservation Authorities Act of 1946 made provision for each Conservation Authority to control drainage works. However, by 1949, these provisions were repealed because one Conservation Authority had been

successful in obtaining a court injunction to prevent the installation of an outlet drain . . . Such controls were in direct conflict with municipal drainage rights to initiate drainage in an intensely cultivated agricultural area (Kettel and Day, 1974, p. 336).

Conservation Authorities were later given powers to request an environmental appraisal of any drainage proposal. The application of this mechanism by the Grand River Conservation Authority was ineffective, in part, because there was an absence of meaningful negotiations among the relevant participants (Shrubsole, 1989).

The implementation of the recommendations of the Review Committee could realize greater economies of scale, improved accountability and increased financial burdens for member municipalities. However, the basic problems confronting the management of water resources would remain because the issues of integrated resource planning and collaboration with line provincial agencies have not been addressed adequately. Any collaboration among public agencies has been *ad hoc* and in the absence of an explicit set of normative goals, values and decision guides. In the past, individual agencies have identified and pursued their own normative, strategic and operational goals without an effective means of resolving legitimate conflicts with other resource users and agencies. Conservation Authorities have been unable to foster an adequate level of communication among these participants.

Effectively aligning the efforts and programmes of water resource agencies is a challenge which Ontario shares with other provinces and the federal government. In Manitoba, Conservation Districts were formed in 1959 to deal with all aspects of soil, water and related resource problems. One of the major impediments confronting these agencies has been the unwillingness of municipalities to relinquish power over drainage works (Carlyle, 1980; 1983; 1984). The merging of the Conservation District and Planning Acts is being contemplated by the Manitoba Government in order to integrate the efforts of Conservation Districts and municipalities more effectively (Canadian Association of Geographers, 1988a).

During the 1970s, some individuals believed that coastal zone management agencies should be established in order to administer fisheries, offshore oil developments, tourism and related onshore infrastructures (Harrison and Parkes, 1983; Johnston *et al.*, 1975; Sadler, 1983). This approach was never implemented, however, because there was a lack of political and bureaucratic support for it (Pross, 1980). The formation of the

Saskatchewan Water Corporation represents an agency which integrates several aspects of water supply (McLean, 1986). Since water issues are interwoven, it must work with other agencies in addressing water quality. Perhaps the development of a water policy could be one method of enhancing co-ordinated efforts among public and private water managers in Canada.

Policy and legislation: the Federal Water Policy (1988)

In 1988, the Federal Government released a water policy which represented a reformulation of federal initiatives (Environment Canada, 1988). Following the publication of the *Report of the Inquiry on Federal Water Policy* in 1985 and the *Report of the National Task Force on Environment and Economy* in 1987, it was believed that a new direction in federal water policy was required because

almost two decades [had] passed since the current direction of federal water policy was established and the legislative base substantially expanded to protect water systems from the adverse impact of a rapidly expanding industrial society. In retrospect, the management approach of the 1970s could be characterized as reactive—responding to and dealing with problems as they arose. This approach has had some success with highly visible forms of pollution and other conventional water issues, but is now proving inadequate (Environment Canada, 1988, p. 1).

The 1988 federal water policy was characterized as being 'anticipatory and preventative' in addressing the management goals of improving water quality and quantity (Environment Canada, 1988, p. 5). These goals were to be realized through three processes and mechanisms: exhortation (i.e., public awareness, science leadership and comprehensive planning), economic incentives (i.e., water pricing) and regulation (i.e., legislation which affects water quality and quantity).

During the 1970s, comprehensive planning was one of the central mechanisms which the Federal Government utilized in the management of water resources. The passage of the Canada Water Act in 1970 legitimized joint federal–provincial planning efforts. It also provided for unilateral federal action if water issues were of significant national interest. On this basis, some 'provinces criticized it vociferously as being cavalier and encroaching upon provincial jurisdiction' (MacLaren, 1972, p. 81). Despite this initial opposition and a continued reluctance by provinces to accept federal participation in provincial matters, a number of comprehensive plans have been completed. Public participation was believed to be an integral part of the planning process. These plans were believed to be important not only in solving regional problems but in the establishment of long-term environmental programmes and pollution control guidelines.

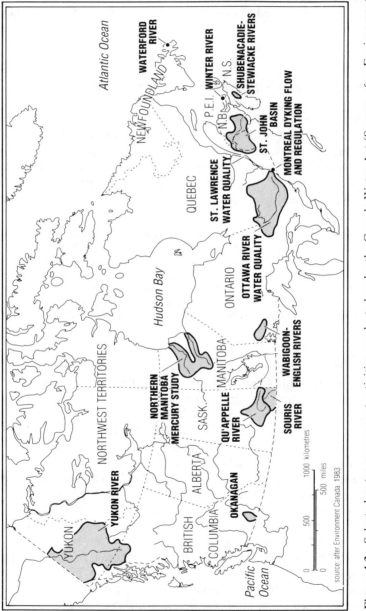

Figure 4.3   Selected water managment activities conducted under the Canada Water Act (Source: after Environment Canada, 1983)

The conduct of studies under the Canada Water Act has evolved through two distinct phases.

Prior to 1975, studies were comprehensive in the treatment of all physical and social processes. Examples include the St. John, Okanagan, Qu'Appelle, Souris and Shebenacadie-Stewiacke basins (see Figure 4.3). While the concept of comprehensive planning might be attractive in theory, it had several shortcomings in practice during the 1970s and 1980s (Brulé et al., 1981; DePape, 1985; Dorcey, 1987a; Mitchell, 1983; 1986b). These problems related to the questionable need for always undertaking a comprehensive plan, the inordinate number of recommendations, the length of time and high cost of completing the plan and the failure to establish an adequate framework for implementation.

The experience of the St. John River basin in New Brunswick illustrates a number of these deficiencies. The study was conducted between 1970 and 1974 at a cost of $912,000, 90 per cent of which was paid by the Federal Government. Water problems relating to industrial and municipal point sources and agricultural diffuse sources were major resource issues to be addressed (Environment Canada, 1975b). The implementation of the plan was also to achieve social and economic improvement of the region which had been plagued continually by comparatively low incomes, high unemployment and underemployment, low rates of labour participation, high net outflow of population and an industrial sector heavily dependent upon the service sector (Fisheries and Environment Canada, 1976). Comprehensive planning was viewed as a means to achieve social, economic and environmental ends.

The final report put forward approximately 150 recommendations. A 1978 review of these found that

[many] had been or were being implemented, that some were too confused or theoretical to be handled and that others were statements of fact, rather than recommendations. As a result, [it was recommended] that agencies should carry out the . . . recommendations as best they could under existing administrative and financial arrangements. Unfortunately, agencies had found that using the Basin Plan as support for a proposal was not necessarily going to have a positive effect. Consequently, implementation of the . . . recommendations has been proceeding slowly and sporadically with ongoing department programs (Cardy, 1983, p. 66).

Thus, implementation had occurred on an *ad hoc* basis through existing institutional arrangements and in the absence of recommendations which identified priorities. Detailed inventories and complex computer simulations provided more information than required, consumed valuable staff and monetary resources and detracted from incorporating fully public input into the final report. According to a senior provincial water manager, 'the process was, in effect, to be the product', rather than the means to achieve the ends (Cardy, 1983, p. 67). The need for a comprehensive plan was also questioned given that, other than water pollution, there was not a serious resource problem. After the writing of the plan, the pollution problem remained and was not solved effectively because of the inadequate implementation framework. This experience is illustrative of the problems associated with comprehensive water planning under the Canada Water Act.

The Federal Government responded to these criticisms through three adjustments. First, plans were undertaken for specific problems in a watershed context. Examples of this integrated planning approach included the Montreal Dyking and Flow Regulation Study (1976), the St. Lawrence Water Quality Planning (1978), the Ottawa River Water Quality Study (1982), the Wabigoon-English Mercury Contamination Study (1984), the Fraser River Estuary Planning (1984) and the Northern Manitoba Mercury Study (1986) (see Figure 4.3). Second, pre-planning agreements were utilized to scope the design of comprehensive studies before they were undertaken. This arrangement was utilized in the Yukon (1979), and Winter basins (1983) (see Figure 4.3). Third, issue specific programmes were introduced under the umbrella of the Canada Water Act. For instance, the Canada Flood Damage Reduction Program was introduced in 1975 and provided for the consideration and implementation of structural and non-structural flood adjustments through joint federal–provincial cost-sharing arrangements (MacLock and Page, 1980). In 1984, the establishment of the Canadian Heritage River System provided for the protection and enhancement of rivers of significant ecological and cultural importance to Canada through co-operative government arrangements. These responses indicated that the Federal Government was pursuing an integrated water management approach.

Some provinces responded to the paralysis which surrounded comprehensive planning by developing their own initiatives. The previously mentioned watershed plans completed by the Conservation Authorities of Ontario would be one example. British Columbia developed a strategic planning system 'to guide long-term policy and management of all natural resources under its jurisdiction' (O'Riordan, 1983a, p. 86). In contrast to the detailed inventories associated with the comprehensive planning of the St. John basin, strategic planning utilized existing data in order to reduce the time required to write the document. A framework for implementation was provided by linking the plan to the budgetary process of the BC Ministry of the Environment (O'Riordan, 1983a; 1983b). A regional approach to river basin planning also was pursued in Alberta. However, water resource planning was seen as a secondary aspect which would facilitate integrated land use planning (Primus, 1983). In 1988, the Government of Alberta proclaimed a Soil Conservation Act which provides a framework to encourage conservation practices through penalties and the clarification of responsibilities. Following the planning experience of the St. John basin, the Province of New Brunswick decided that future approaches would be flexible and 'allow for the appropriate control and regulation at the time and place it [was] required' (Cardy, 1983, p. 69).

The Federal Water Policy document reaffirms the tradition of comprehensive water planning which 'takes into account all water uses and water-related activities' on a watershed basis (Environment Canada, 1988, p. 10). While the pursuit of this approach to water resource planning may be intuitively attractive, the recent initiatives of Federal and Provincial Governments have indicated that integrated efforts should dominate the practice of management in Canada.

A second emphasis of the Federal Water Policy was the control of toxic substances. The 1985 Inquiry on Federal Water Policy criticized arrangements surrounding the management of toxic substances because they placed the onus on government, rather than on the manufacturer, to identify contaminants. In addition, only five substances had been scheduled under the Environmental Contaminants Act since its inception in 1975 and co-operation among line federal and provincial departments was ineffective (Pearse *et al.*, 1985; Pearse, 1986).

In 1988, these shortcomings were addressed. The passage of the Canadian Environmental Protection Act shifted the onus from government to the manufacturer in documenting the chemical properties and environmental and health effects associated with chemical substances. This proactive stance was complemented by provisions which allowed for regulations at all stages—development, manufacture, storage, transport, use and disposal—of any scheduled substance. Citizen rights were enhanced and penalties for non-compliance were increased (Environment Canada, May 1988). One weakness of the legislation is related to the fact that only nine substances are regulated presently.

Since responsibility for water quality is shared with the provinces, specific agreements can be negotiated which allow equivalent provincial regulations to supercede federal restrictions. Thus, the regulated substances identified in the Canadian Environmental Protection Act represent the lowest form of protection. All provincial governments have developed their own laws for dealing with chemical and hazardous products. In Ontario, a Municipal–Industrial Strategy for Abatement (MISA) was initiated in order to address this issue (Ontario Ministry of the Environment, March 1988).

The third thrust of the Federal Water Policy related to water pricing. The Inquiry on Federal Water Policy concluded that existing pricing mechanisms were rudimentary. Furthermore, the perceived abundance of water in Canada has encouraged Canadians to take it for granted. According to Environment Canada (1983, p. 10), 'traditionally, water has been considered a free good, and charges for its use have been related to the costs of treatment prior to use, distribution and pollution abatement. In contrast to most other natural resources, no commodity charge is attached to water itself.' Since there is a lack of economic incentives to promote efficient use, Canadians have not effectively managed their use patterns. The Federal Water Policy advocated 'a fair value' be charged for water in order to encourage efficient use and to provide increased funds for the investment into the water infrastructure (Environment Canada, 1988, p. 8).

There has not yet been any significant movement toward the pricing of water. Since this issue is primarily a provincial responsibility, negotiations between these levels of government are necessary. One obstacle which confronts these discussions is the potential ramifications for regional development.

The Federal Water Policy emphasizes the need for federal–provincial negotiation in the planning of river basins and in the control of chemical substances. Federal participation in these activities is legitimized through cost-sharing arrangements and legislation. Communication between these levels of government is one important aspect of an effective water policy for Canada. However, the document fails to address the commitment of the Federal Government to this initiative. Hopefully, it will not experience the difficulties encountered in the implementation of the Canada Water Act. Weaknesses in the Canadian economy and the need to cut the federal budget deficit in the late 1970s and 1980s, combined with the disenchantment surrounding comprehensive river basin planning, contributed to fewer financial resources being allocated to environmental programmes (Brulé et al., 1981; Dorcey, 1987a). Without an adequate commitment to fund water programmes by the federal and provincial governments, the effective implementation of the Federal Water Policy will be difficult. The experience of the Canada Water Act indicates that water management must be placed in the context of the total political agenda (i.e., social, economic, environmental) and must compete for funds with other government programmes. There is a need to consider water resource problems with other issues.

Environmental Impact Assessment and normative planning efforts

The 1970s witnessed an increase in the development of environmental policies as governments responded to heightened public concern for ecological and social values. Evidence of this concern was seen, in part, in the development of provisions for environmental impact assessment (EIA) by federal and provincial governments. Mitchell and Turkheim (1977, p. 47) defined EIA as

an activity designed to identify, interpret and communicate information about the impact of an action on [human] health and well-being (including the well-being of ecosystems on which [human] survival depends).

In this context, impact assessment should be broad in scope and should examine direct and indirect effects, both positive and negative, on human and non-human environments. The use of EIA is one method to consider water in the context of other social, environmental and economic issues.

Government departments responsible for environmental impact assessment processes derive their authority through three mechanisms (Couch, 1988). Saskatchewan, Ontario and Newfoundland have passed individual Environmental Impact Assessment Acts. British Columbia, Alberta, Quebec, Nova Scotia and Prince Edward Island have integrated provisions for environmental impact assessment into existing water, land and/or utility based legislation. A third option which is utilized by Manitoba, New Brunswick and Canada is to adopt a policy for environmental impact assessment.

The need for coastal zone management has been associated with the increased demands placed upon the resource base by a variety of competing user groups, such as fisheries and offshore oil developments. Environmental impact assessment would be one mechanism to determine the feasibility and desirability of offshore energy resource development. The management of these resources is the responsibility of the Federal Government and would be subject to the requirements of the Federal Environmental Assessment Review Process (EARP).

The EARP was established in 1973 to provide for the assessment of environmental consequences of any proposal undertaken directly by all federal departments and crown agencies, or that may have an environmental effect on an area of federal responsibility. It is conducted in two phases: environmental screening by the initiating department to determine if the proposal is likely to have significant environmental effects, and formal environmental assessment by the Federal Environmental Assessment and Review Office (FEARO) (see Figure 4.4a). If, after self-assessment, a proponent determines that there may be significant environmental consequences, an assessment panel is formed by FEARO to review publicly the proposal. An impact statement is prepared by the proponent and is to meet the requirements established by the panel. Between 1972 and 1987, only 32 reviews have been held (FEARO, 1987). This indicates the crucial importance of the screening stage (see Figure 4.4a). Reports are prepared for the Minister of the Environment for those proposals which are reviewed by the panel. The Minister then submits a recommendation to the Minister of the initiating department. Should significant difference of opinion arise, Cabinet is utilized as the final decision-making body (see Figure 4.4a).

Rees (1980) and King and Nelson (1983) examined the effectiveness of the EARP process. According to the overview study by Rees (1980), and an examination of an offshore oil development in the South Davis Strait by King and Nelson (1983), EARP has failed to ensure an adequate level of environmental management because of inadequate and inconsistent

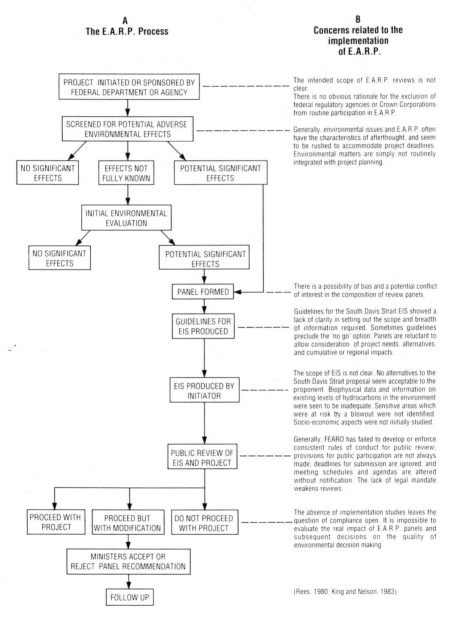

Figure 4.4   The Federal Environmental Assessment Review Process: theory and practice (Source: Rees, 1980; King and Nelson, 1983)

implementation (see Figure 4.4b). In 1988, the Federal Minister of the Environment announced his intention of strengthening EARP by upgrading its present policy status to a legislative act (Canadian Association of Geographers, 1988b). This development would enhance the legitimacy of the process because all activities of the Federal Government would require environmental assessments and there would be less importance placed on self-screening.

A second issue in the administration of EARP relates to division of federal–provincial responsibilities. During the 1980s, the Government of Saskatchewan emphasized the development of water-based resources as a means of enhancing the economic diversification of the province. This was highlighted in the formation of the Saskatchewan Water Corporation which is responsible for a range of water supply, treatment and development services (McLean, 1986). One of the projects supported by the Corporation was the construction of two reservoirs, the Rafferty and the Alameda, in the Souris River Valley in order to provide cooling water to coal-fired electrical generators, adequate water supply and flood protection (Souris Basin Development Authority, 4 August 1987). While these goals are the prime responsibility of the provincial government, the Federal Government is involved in the management activities of the Souris River since it is an international and interprovincial river (see Figure 4.3).

In 1986, the Souris Basin Development Authority was formed to plan and build the facilities. Approval of the project was required from the Saskatchewan Water Corporation under the Saskatchewan Water Corporation Act, the Saskatchewan Minister of the Environment and Public Safety and Environment Canada under the International River Improvements Act. To receive building and operating licences from these agencies, the Development Authority completed an Environmental Impact Statement in August 1987 (Souris Basin Development Authority, 1987). By June 1988, the Development Authority had received the necessary permits and licences from the provincial agencies and Environment Canada.

During the approval process, the Canadian Wildlife Federation, a nongovernmental organization, had lobbied Environment Canada to submit the project to EARP because (1) the negative impacts on wildlife and wildlife habitat were substantial and significant and (2) the Environmental Impact Statement did not examine impacts in Manitoba or the United States. When their attempt to invoke the EARP failed and Environment Canada issued the licence, the Canadian Wildlife Federation appealed to the Federal Court of Canada. The Canadian Wildlife Federation maintained that Environment Canada must comply with the provisions of EARP before granting approval under the International River Improvements Act. By not doing so, the Minister had exceeded his jurisdiction.

Environment Canada believed that it was inappropriate to apply the EARP guidelines to the Rafferty/Alameda project since it was a provincial initiative which was funded by the province of Saskatchewan and had been subjected to a formal environmental review and board of inquiry by the Saskatchewan Department of the Environment and Public Safety. To undertake a federal review of the project, which had already been scrutinized by the Saskatchewan process, and which in principle met the requirements of EARP, would be an unwarranted duplication of effort.

The Federal Court ruled in support of the position of the Canadian Wildlife Federation. The Court noted that the Environmental Impact Statement had not adequately addressed federal concerns related to post-project monitoring, water quality and quantity issues, impacts on wildlife and habitat and the protection of navigable waters. The statement had also dealt exclusively with impacts within the province and omitted a review of impacts in Manitoba and the United States. The Court maintained that the application of EARP to this project would not represent undesirable duplication since the shortcomings of the provincial study would be addressed. In addition, the Court ruled that Environment Canada had not complied with the provisions of EARP and must do so since the project would impact on several areas of federal responsibility. This judgement highlights the need for effective scoping and communication among all relevant participants during the conduct of environmental impact assessments.

Impact assessment could also be a useful technique in appraising the desirability and feasibility of interbasin transfers. At present, there are over 50 diversions in Canada, and there have been considerations concerning future projects which include transfers from northern to southern Alberta, and from James Bay to the Great Lakes (Bocking, 1987; Rosenberg et al., 1987). The experience of the Kemano Diversion in British Columbia provides further insight as to the effectiveness of environmental assessment in Canada.

In 1950, the Government of British Columbia permitted the Aluminum Company of Canada (Alcan) to develop a hydroelectric generating facility and to divert water from the Nechako River into the Nanika River in order to develop an aluminium smelter (Day, 1985). Water rights were guaranteed in perpetuity if Alcan achieved specific capacity requirements by 1983 (Gomez-Amaral and Day, 1987). During the late 1970s, decreased flows on the Nechako River were particularly detrimental to fish populations as Alcan diverted more water in order to increase electrical production. This prompted the Federal Department of Fisheries and Oceans, which has jurisdiction over fisheries, to seek and obtain a court injunction which restricted the amount of water which could be diverted.

Between 1980 and 1987, there were heated discussions regarding the perceived intervention of the Federal Government into provincial responsibilities, the arbitrary nature of the flow schedule and the negative economic aspects of this injunction. After protracted arguments in court and several conflicting scientific studies concerning the relationship between flow and fish populations, an out-of-court agreement was achieved among Alcan, the Provincial Ministry of Energy, Mines and Petroleum Resources and the Federal Department of Fisheries and Oceans. This arrangement called for increased protection to fish populations through the implementation of a revised flow schedule, increased generating capacity for Alcan and obliged the provincial government to mitigate against any negative impacts on resident fish populations (Gomez-Amaral and Day, 1987). While this agreement serves multiple purposes, it neglects the interests of other resource users. Gomez-Amaral and Day (1987, pp. 148–9) commented on the decision-making process by noting that

the federal government could have used the Federal Environmental Assessment and Review Process, or the province could have used its Energy Review Process to appraise the agreement in public. As a result, the fishermen's union and the Rivers Defence coalition denounced the agreement and have called for a more open decision-making process to decide the uses and the allocation of waters in all provincial rivers. They may also challenge the legality of the agreement in the courts.

In addition, the Cheslatta Indian Band which was affected along the spillway route may pursue a legal challenge in order to regain their losses (Day, 1985). In the 1950s, its land was flooded for seven months before the Federal Minister of Indian Affairs approved the sale to Alcan for $125,000.

While environmental impact assessment is a tool which can facilitate effective water resource development, the experiences of EARP and the Kemano River suggest that its implementation in Canada requires improvement. At present, important planning is done behind closed doors and has made the public suspicious about governments, large industries and major water projects (Day and Quinn, 1987). For instance, efforts to reform EARP in 1984 were frustrated by inter-agency rivalries which softened or eliminated desired improvements (Fenge and Smith, 1986). There are two lessons to be learned from these experiences. First, impact assessment must be applied on a more consistent basis. At present, the need for review is not well established or else is left to the discretion of the responsible Minister. Longer lead times are required and managers must be more sensitive to the needs of affected groups, particularly natives. Second, impact assessment has dealt primarily with operational issues and not addressed strategic or normative questions. The latter questions would examine the impact of policies and programmes while the former is restricted to the consideration of project impacts. An exception to this generalization would be the Berger Commission Inquiry into the MacKenzie Valley Pipeline (Berger, 1977).

The incorporation of water, as well as other environmental and social issues, into the policy making process at a normative level would dramatically alter the past practice of water management which has overvalued economic considerations and structural solutions (Smith, 1982). The need for this level of discussion, which related directly to the management of the water resources of Canada, was demonstrated during the federal election of 1988 which focused on the issue of a 'Free Trade Deal' with the United States.

In September 1985, the Federal Government announced that it would seek a free trade agreement with the United States in order to provide for increased economic benefits to both countries through the removal of trade barriers. In 1988, a final agreement was negotiated and awaited ratification by the federal parliament when an election was held. Much of the election focused on the social, economic and environmental implications of this arrangement. Some individuals felt that water was to be treated as an economic good in the deal and would be subject to massive export to the United States through interbasin transfers (Clark and Gamble, 1988; Holm, 1988). Although this assertion was denied by the Government, Ontario responded to this perceived threat to its water resources by introducing legislation which would regulate large-scale water exports (Danson, 1988). Following the re-election of the Federal Government, the Free Trade Agreement was ratified.

This debate highlights the need to consider the environmental, social and economic consequences of resource development decisions at normative and strategic policy levels. There have been three initiatives which have occurred in the 1980s which could provide this level of debate.

First, the writing of Conservation Strategies has been undertaken by federal and provincial governments. This development followed the lead of the World Conservation Strategy which stressed the need for the conservation, maintenance and wise management of versatile and productive ecosystems as a basis for social and economic growth (IUCN, 1980). In a review of Canadian and international experiences with the writing of National Conservation Strategies (NCS), Nelson (1987, pp. 48–9) observed that

many people see the NCS process as a comprehensive and wide ranging one which requires much information as a basis for planning for major changes in the future . . . Yet concern about the process, contents, and priorities, and other *essentially political* aspects of a NCS inevitably will raise questions about how it has been prepared, by whom and for whom. There will also be questions about what its relationship is to other decision-making documents and to planning and political procedures, notably at provincial and local levels.

The objectives of a Conservation Strategy are to maintain ecological processes, to preserve genetic diversity and to ensure the sustainable utilization of species and ecosystems (IUCN, 1980). Nelson (1987) concluded that the writing of the documents was being conducted by professionals in the absence of extensive public participation. There was a need to broaden not only the participant base but also the considerations of Conservation Strategies.

The formation of 'Round Tables' attempts to address the weaknesses of the Conservation Strategies. Following the publication of *Our Common Future*, or the Brundtland Report, in 1987, the Canadian Council of Resource and Environment Ministers established the National Task Force on Environment and Economy 'to look at environment-economy integration from a Canadian perspective (National Task Force on Environment and Economy, 1988, p. 1). The formation of multi-sectoral Round Tables was a key recommendation of this body. Comprising senior decision-makers from industry and labour, non-governmental organizations, academics and politicians, the Round Tables are intended to provide a forum for senior representatives from many sectors of Canadian life to express their views concerning sustainable development, and to develop innovative approaches to address environment, economy and society linkages. The federal and seven provincial governments have formed Round Tables (National Task Force on Environment and Economy, 1988). Following the election of 1988, the Federal Government announced that the Minister of the Environment would sit at the important Planning and Priorities Committee of Cabinet in order that environmental issues are given the same treatment as economic considerations.

The formation of Remedial Action Plans (RAPs) groups in the Great Lakes Basin is another mechanism for the public to participate in water management. These groups are designed to define the necessary actions and timetables and restore water quality in 42 areas of concern (Hartig, 1987). Since this is an informal process which is not legitimized through legislation, momentum acquired during the planning phase must drive the commitment to allocate the funds necessary for implementation. This process is in contrast to the public hearings which must be held during the licensing process by the Water Boards established by the Northern Inland Waters Act. Water licensing in the north is very much a public process, whereby northern residents who may be affected by a given water use voice their concerns (Rueggeburg, 1985, p. 31). Indeed, a variety of public forums could provide for more effective water management in Canada.

Implications

Canada has not adopted a single solution to remedy water management problems. Canadian federalism discourages a single national perspective and supports a belief that most water management problems are regional in nature. The absence of a national perspective is reinforced in the Constitution which identifies the ten provinces as the primary water managers for Canada. The variety of water issues and the emphasis on negotiated agreements among the levels of government has meant that institutional arrangements have been tailored to address the physical and human circumstances of a region. Rather than adopting a uniform approach to problems, institutional arrangements for the management of water resources resemble a patchwork quilt.

The regional perspective of water resource problems and solutions could be altered if large-scale water exports are considered in the future. The impact of these actions would have implications for the environment, economy and society at a national level. This regional perception could also be altered if the efforts of the Round Tables are successful in identifying how water development contributes to national and provincial economic and social planning. Although a single correct approach is not to be found in Canada, there are a number of lessons which may be drawn from this overview.

First, existing legislation and regulatory mechanisms are adequate to legitimize the status of water management agencies. This legal basis is supported primarily through regulatory mechanisms and cost-sharing arrangements. The latter has been an important tool in the practice of federalism. Financial arrangements among federal, provincial and local levels of government have facilitated the consideration of the interests of all levels of governments.

Second, water management functions are dispersed not only among levels of government, but also within levels of government. Comprehensive water management in Canada is achieved through the combined efforts of a number of integrated water management agencies.

There have been two general approaches to foster effective co-operation among these participants. The use of comprehensive planning as seen through the Conservation Reports and watershed plans of the Conservation Authorities, the plans completed under the Canada Water Act and impact statements under environmental legislation have addressed multiple resource use conflicts. The Federal Water Policy advocates that this tradition be continued and strengthened. However, the shortcoming of this co-operative approach is that

it is a very unusual situation in which conflicts between uses do not arise. It is equally unusual to find effective mechanisms and decision-making criteria for conflict resolution. Until such mechanisms and criteria are developed and used, good water resource management will be difficult if not impossible (Thompson, 1987, p. 9).

An improvement in the practice of public participation at normative, strategic, and operational levels of decision making would address these concerns.

Another method of finding effective conflict resolution mechanisms and decision criteria is to challenge the traditional view that effective water management in Canada is a co-operative approach. While the efforts of more than one level of government are often required because of the complicated jurisdictional arrangements, 'co-operation' may not be an accurate description of water management in Canada. Perhaps it would be appropriate to focus on the nature of achieving inter-agency and inter-governmental agreements. This approach was supported by Thompson (1980) and Dorcey (1987b) who maintained that bargaining, not co-operation, is an integral, yet poorly understood process in the management of Canadian water resources. This approach to the study of institutional arrangements would focus more attention on the organizational culture and participant attitudes.

This view of bargaining is consistent with Saunders' (1986, p. xi) contention that the purpose of Canadian federalism is not to eliminate conflict in that, 'the very nature of federalism presupposes that a diversity in political responses to a given issue is a good thing'. Healthy conflict situations and continuous discussion among all relevant participants, public and private, are key elements in achieving effective agreements from this perspective of federalism. In essence, it is a bargaining process.

Increasing demands, complexity and uncertainty are likely to make water resource management in Canada more challenging (Dorcey, 1987b). This trend is evident in the discussions concerning the impact changes in climate could have on the water management (Hengeveld, 1987). This is one area of research which challenges both physical and social scientists.

Numerous weaknesses associated with water management in Canada have been identified. There are signs of significant opportunities to improve management as evidenced by the formation of Round Tables. Improvements should focus on the organizational culture and participant attitudes, and the mechanisms and processes. Prescriptions should better incorporate the views of all relevant participants into the management process and should ensure that a broad range of alternatives is rigorously evaluated. Perhaps the greatest challenge facing those in water management is to indicate how individuals can better contribute and to identify how management efforts can be sustained and perhaps enhanced during periods of fiscal restraint. If the commitment to water management is real, then the public must be willing to pay for and governments be willing to fund programmes.

References

Barton, B., 1986, 'Cooperative management of interprovincial water resources' in J. O Saunders (ed.), *Managing Natural Resources in a Federal State*, Carswell, Toronto, Ontario, pp. 235–50.

Berger, T. R., 1977, *Northern Frontier, Northern Homeland* (2 vols.), James Lorimer & Company, Toronto, Ontario.

Bird, P. M. and Rapport, D. J., May 1986, *State of the Environment Report for Canada*, Minister of Supply and Services Canada, Ottawa, Ontario.

Bocking, R. C., 1987, 'Canadian water: A commodity for export?' in M. C. Healey and R. R. Wallace (eds.),*Canadian Aquatic Resources*, Minister of Supply and Services Canada, Ottawa, Ontario, pp. 105–35.

Brulé, B., Quinn, F., Wiebe, J. and Mitchell, B., 1981, *An Evaluation of the River Planning and Implementation Programs*, Inland Waters Directorate, Environmental Conservation Service, Planning and Evaluation Directorate, Corporate Planning Group, Environment Canada, Ottawa, Ontario.

Burton, T. L., 1972, *Natural Resource Policy in Canada: Issues and Perspectives*, McClelland and Stewart, Toronto, Ontario.

Cairns, R. D., 1987, 'An economic assessment of the resource amendment', *Canadian Public Policy*, **XIII** (4), pp. 504–14.

Canadian Association of Geographers, 1988a, 'Land and water management in Manitoba', *The Operational Geographer*, **6** (3), p. 68.

Canadian Association of Geographers, 1988b, 'Federal environmental assessment and review process strengthened', *The Operational Geographer*, **6** (3), p. 66.

Cardy, W. F. G., 1983, 'River basins and water management in New Brunswick' in B. Mitchell and J. S. Gardner (eds.), *River Basin Management: Canadian Experiences*, Department of Geography Publication Series no. 20, University of Waterloo, Waterloo, Ontario, pp. 61–70.

Carlyle, W. J., 1980, 'The management of environmental problems on the Manitoba escarpment', *The Canadian Geographer*, **XXIV** (3), pp. 255–69.

Carlyle, W. J., 1983, 'Agricultural drainage in Manitoba: the search for administrative boundaries' in B. Mitchell and J. S. Gardner (eds.), *River Basin Management: Canadian Experiences*, Department of Geography Publication Series no. 20, University of Waterloo, Waterloo, Ontario, pp. 279–96.

Carlyle, W. J., 1984, 'Water in the Red River Valley of the North', *Geographical Review*, **74** (3), pp. 331–58.

Cherry, J. A., 1987, 'Groundwater occurrence and contamination in Canada' in M. C. Healey and R. R. Wallace (eds.), *Canadian Aquatic Resources*, Minister of Supply and Services Canada, Ottawa, Ontario, pp. 307–26.

Clark, M. and Gamble, D., 1988, 'Water is in the deal' in W. Holm (ed.), *Water and Free Trade: the Mulroney Government's Agenda for Canada's Most Precious Resource*, James Lorimer & Company, Toronto, Ontario, pp. 2–20.

Couch, W. J., 1988, *Environmental Assessment in Canada: 1988 Summary of Current Practice*, Canadian Council of Resource and Environment Ministers, Minister of Supply and Services, Ottawa, Canada.

Danson, T., 1988, 'A federal–provincial minefield' in W. Holm (ed.), *Water and Free Trade: the Mulroney Government's Agenda for Canada's Most Precious Resource*, James Lorimer & Company, Toronto, Ontario, pp. 122–8.

Day, J. C., 1985, *Canadian Interbasin Diversions*, Inquiry on Federal Water Policy, Research Paper no. 6, Ottawa, Ontario.

Day, J. C. and Quinn, F., 1987, 'Dams and diversions: learning from the Canadian experience' in W. Nicholaichuk and F. Quinn (eds.), *Proceedings of the Symposium on Interbasin Transfer of Water: Impacts and Research Needs for Canada*, Environment Canada, Ottawa, Ontario, pp. 43–58.

DePape, D., 1985, *Federal–Provincial Co-operation in Water—an Exploratory Examination*, Inquiry on Federal Water Policy, Research Paper no. 9, Ottawa, Ontario.

Doern, G. B. and Phidd, R. W., 1983, *Canadian Public Policy: Ideas, Structure, Process*, Methuen, Agincourt, Ontario.

Dorcey, A. H. J., 1987a, 'Research for water resources management: the rise and fall of great expectations' in M. C. Healey and R. R. Wallace (eds.), *Canadian*

Aquatic Resources, Minister of Supply and Services, Ottawa, Ontario, pp. 481–511.

Dorcey, A. H. J., 1987b, 'The myth of interagency cooperation in water resources management', *Canadian Water Resources Journal*, **12** (2), pp. 17–26.

Environment Canada, 1975a, *Canada Water Year Book 1975*, Information Canada, Ottawa, Ontario.

Environment Canada, 1975b, *Monograph on Comprehensive River Basin Planning*, Information Canada, Ottawa, Ontario.

Environment Canada, 1983, *Canada Water Year Book 1981–1982; Water and the Economy*, Minister of Supply and Services Canada, Ottawa, Ontario.

Environment Canada, 1988, *Federal Water Policy*, Environment Canada, Ottawa, Ontario.

Environment Canada, May 1988, *Canadian Environmental Protection Act Enforcement and Compliance Policy*, Minister of Supply and Services Canada, Ottawa, Ontario.

Estrin, D. and Swaigen, J., 1978, *Environment on Trial* (revised edn.), Canadian Environmental Law Research Foundation, Toronto, Ontario.

FEARO, 1979, *Revised Guide to the Federal Environmental Assessment and Review Process*, Minister of Supply and Services Canada, Ottawa, Ontario.

FEARO, 1987, *Register of Panel Projects*, Number 24, Minister of Supply and Services Canada, Ottawa, Ontario.

Fenge, T. and Smith, L. G., 1986, 'Reforming the federal environmental assessment and review process', *Canadian Public Policy*, **XII** (4), pp. 596–605.

Fisheries and Environment Canada, 1976, *Canada Water Year Book 1976*, Minister of Supply and Services Canada, Ottawa, Ontario.

Foster, H. D. and Sewell, W. R. D., 1981, *Water: the Emerging Crisis in Canada*, James Lorimer & Company, publishers in association with The Canadian Institute for Economic Policy, Toronto, Ontario.

Gomez-Amaral, J. C. and Day, J. C., 1987, 'The Kemano diversion: a hindsight assessment' in W. Nicholaichuk and F. Quinn (eds.), *Proceedings of the Symposium on Interbasin Transfer of Water: Impacts and Research Needs for Canada*, Environment Canada, Ottawa, Ontario, pp. 137–52.

Harrison, P. and Parkes, J. G. M., 1983, 'Coastal zone management in Canada', *Coastal Zone Management Journal*, **11** (1–2), pp. 1–11.

Hartig, J., 1987, 'Protocol established for review of Remedial Action Plans for areas of concern', *Focus*, **12** (I), p. 7.

Hengeveld, H. G., 1987, 'Climate change and water resources' in W. Nicholaichuk and F. Quinn (eds.), *Proceedings of the Symposium on Interbasin Transfer of Water: Impacts and Research Needs for Canada*, Environment Canada, Ottawa, Ontario, pp. 25–26.

Holm, W. R., 1988, 'The pressure to sell out water' in W. Holm (ed.), *Water and Free Trade: the Mulroney Government's Agenda for Canada's Most Precious Resource*, James Lorimer & Company, Toronto, Ontario, pp. 28–43.

IUCN, 1980, *World Conservation Strategy, Living Resource Conservation for Sustainable Development*, prepared with the United Nations Environment Programme, the World Wildlife Fund, the Food and Agricultural Organization and the United Nations Educational, Scientific and Cultural Organization, Gland, Switzerland.

Johnston, D. M., Pross, A. P., McDougall, I. with Dale, N. G., 1975, *Coastal Zone*

Framework for Management in Atlantic Canada, Dalhousie Institute of Public Affairs for Environment Canada, Halifax, Nova Scotia.

Kettel, G. A. and Day, J. C., 1974, 'Outlet drainage in Ontario: a methodological exploration', *Journal of Environmental Management*, **2**, pp. 331–49.

King, J. E. and Nelson, J. G., 1983, 'Evaluating the federal environmental assessment and review process with special reference to South Davis Strait, northeastern Canada', *Environmental Conservation*, **10** (4), pp. 293–301.

Lucas, A. R., 1986, 'Harmonization of federal and provincial environmental policies: the changing legal and policy framework' in J. O. Saunders (ed.), *Managing Natural Resources in a Federal State*, Carswell, Toronto, Ontario, pp. 33–52.

MacLaren, J. W., 1972, 'Canada's water act', *Journal of Soil and Water Conservation*, March–April, pp. 80–2.

MacLock, R. B. and Page, G. A., 1980, 'Cutting our flood losses' in W. R. D. Sewell and M. L. Barker (eds.), *Water Problems and Policies*, Cornett Occasional Papers no. 1, Department of Geography, University of Victoria, Victoria, British Columbia, pp. 7–12.

McDonald, A. T. and Kay, D., 1988, *Water Resources: Issues and Strategies*, Longman, London.

McLean, R. A., 1986, 'Saskatchewan water corporation', *Canadian Water Resources Journal*, **11** (3), pp. 62–8.

Mitchell, B., 1983, 'Comprehensive river basin planning: problems and opportunities', *Water International*, **8**, pp. 146–53.

Mitchell, B., 1986a, 'The evolution of integrated resource management' in R. Lang (ed.), *Integrated Approaches to Resource Planning and Management*, the University of Calgary Press, Calgary, Alberta, pp. 13–26.

Mitchell, B., 1986b, *Integrated River Basin Management: Canadian Experiences*, Hydrology and Water Resources Symposium, Griffith University, Brisbane, Australia.

Mitchell, B. and Turkheim, R., 1977, 'Environmental impact assessment: principles, practices and Canadian experiences' in R. R. Kreuger and B. Mitchell (eds.), *Managing Canada's Renewable Resources*, Methuen, Toronto, Ontario, pp. 47–66.

Muller, R. A., 1985, *The Socio-economic Value of Water in Canada*, Inquiry on Federal Water Policy, Research Paper no. 5, Ottawa, Ontario.

National Task Force on Environment and Economy, 1988, *Progress Report*, National Task Force on Environment and Economy, Ottawa, Ontario.

Nelson, J. G., 1987, 'Experiences with national conservation strategies: lessons for Canada', *Alternatives*, **15** (1), pp. 42–9.

Ontario Ministry of the Environment, March 1988, *MISA: the Effluent Monitoring Pollutants List (1987)*, Ontario Ministry of the Environment, Toronto, Ontario.

Ontario Ministry of Natural Resources, December 1987, *A Review of the Conservation Authorities Program*, Ontario Ministry of Natural Resources, Toronto, Ontario.

O'Riordan, J., 1983a, 'Strategic planning and river basin management' in B. Sadler (ed.), *Water Policy for Western Canada: the Issues of the Eighties*, the University of Calgary Press, Calgary, Alberta, pp. 86–108.

O'Riordan, J., 1983b, 'New strategies for water resource planning in British Columbia' in B. Mitchell and J. S. Gardner (eds.), *River Basin Management:*

Canadian Experiences, Department of Geography Publication Series no. 20, University of Waterloo, Waterloo, Ontario, pp. 17–42.

Pearse, P. H., 1986, 'Developments in Canada's water policy', in *The Management of Water Resources: Proceedings of an International Seminar*, the Institute for Research on Public Policy and the Institute of Public Administration of Canada, Toronto, pp. 1–18.

Pearse, P. H., Bertrand, F. and MacLaren, J. W., September, 1985, *Currents of Change, Final Report of the Inquiry on Federal Water Policy*, Environment Canada, Ottawa, Ontario.

Powell, J. R., 1983, 'River basin management in Ontario' in B. Mitchell and J. S. Gardner (eds.), *River Basin Management: Canadian Experiences*, Department of Geography Publication Series no. 20, University of Waterloo, Waterloo, Ontario, pp. 49–60.

Primus, C. L., 1983, 'River basin planning in Alberta: current status—future issues' in B. Mitchell and J. S. Gardner (eds.), *River Basin Management: Canadian Experiences*, Department of Geography Publication Series no. 20, University of Waterloo, Waterloo, Ontario, pp. 43–8.

Pross, A. P., 1980, 'The coastal zone management debate: putting an issue on the public agenda' in O. P. Dwivedi (ed.), *Resources and the Environment: Policy Perspectives for Canada*, McClelland and Stewart Ltd, Toronto, Ontario, pp. 107–32.

Quinn, F. 1977, 'Notes for a national water policy' in R. R. Krueger and B. Mitchell (eds.), *Managing Canada's Renewable Resources*, Methuen, Toronto, Ontario, pp. 226–38.

Quinn, F., 1986, 'The evolution of federal water policy', *Canadian Water Resources Journal*, **10** (4), pp. 21–33.

Rees, W. E., 1980, 'EARP at the crossroads: environmental assessment in Canada', *Environmental Impact Assessment Review*, **1** (4), pp. 355–79.

Richardson, A. H., 1974, *Conservation by the People*, University of Toronto Press, Toronto, Ontario.

Romanow, R. J., 1986, 'Natural resources management in a federal context: an overview' in J. O. Saunders (ed.), *Managing Natural Resources in a Federal State*, Carswell, Toronto, pp. 1–14.

Rosenberg, D. M., Bodaly, R. A., Hecky, R. E. and R. W. Newbury, 1987, 'The environmental assessment of hydroelectric impoundments and diversions in Canada' in M. C. Healey and R. R. Wallace (eds.), *Canadian Aquatic Resources*, Minister of Supply and Services Canada, Ottawa, pp. 71–104.

Rueggeberg, H., 1985, *Northern Water Issues*, Inquiry on Federal Water Policy, Research Paper no. 12, Ottawa, Ontario.

Rueggeberg, H. and Thompson, A., 1985, *Water Law in Canada*, Inquiry on Federal Water Policy, Research Paper no. 1, Ottawa, Ontario.

Sadler, B. (ed.), 1983, *Coastal Zone Management in British Columbia*, Cornett Occasional Papers no. 3, Department of Geography, University of Victoria, Victoria, British Columbia.

Saunders, J. O., 1986, 'Good federalism, bad federalism: managing our natural resources' in J. O. Saunders (ed.), *Managing Natural Resources in a Federal State*, Carswell, Toronto, Ontario, pp. ix–xviii.

Saunders, J. O., 1987, 'Management and diversion of interjurisdictional rivers in Canada: a legal perspective', in W. Nicholaichuk and F. Quinn (eds.), *Proceed-*

ings of the Symposium on Interbasin Transfer of Water: Impacts and Research Needs for Canada, Environment Canada, Ottawa, Ontario, pp. 465–79.

Science Council of Canada, June 1988, Water 2020: Sustainable Use for Water in the 21st Century, Science Council of Canada Report 40, Minister of Supply and Services, Ottawa, Ontario.

Shrubsole, D. A., 1989, The Evolution of Integrated Water Management in Ontario: the Cases of the Ganaraska Region and Grand River Conservation Authorities, unpublished Ph.D. dissertation, Department of Geography, University of Waterloo, Waterloo, Ontario.

Smith, L. G., 1982, 'Mechanisms for public participation at a normative planning level in Canada', Canadian Public Policy, VIII (4), pp. 561–72.

Souris Basin Development Authority, 1987, Rafferty–Alameda Project Environmental Impact Statement, 17 chapters, Souris Basin Development Authority, Regina, Saskatchewan.

Souris Basin Development Authority, 4 August 1987, Executive Summary, Rafferty/Alameda Environmental Impact Statement, Souris Basin Development Authority, Regina, Saskatchewan.

Tate, D., 1981, 'River basin development in Canada' in B. Mitchell and W. R. D. Sewell (eds.), Canadian Resource Policies: Problems and Prospects, Methuen, Toronto, Ontario, pp. 151–79.

Thompson, A. R., 1980, Environmental Regulation in Canada, Westwater Research Centre, Vancouver, British Columbia.

Thompson, D., 1987, 'Moving with the currents of change', Canadian Water Resources Journal, 12 (1), pp. 6–15.

Whillans, T. H., 1987, 'Wetlands and aquatic resources' in M. C. Healey and R. R. Wallace (eds.), Canadian Aquatic Resources, Minister of Supply and Services Canada, Ottawa, pp. 321–56.

CHAPTER 5

Integrated water management in England

Alan S. Pitkethly

Introduction

The system for water management in England stands on the threshold of radical change for the third time in three decades. What follows is largely an historical account of three phases of change which might be called 'the age of technical supremacy', 'the age of the accountant' and 'reorientation to market principles'. It is preceded by a discussion of the English view of integration and then followed by an examination of factors that seem to have shaped this view. In the interests of clarity and brevity the differing detail of developments regarding tidal estuaries and coastal waters is omitted as are the special considerations and factors affecting Wales. This is a discussion of water management in England. Co-ordinated management of land and water is an idea that continues to engage serious attention around the world but that has failed to be translated into practice. More realistic is Mitchell's (1987, p. 24) suggestion that at the operational level an integrated approach can work by concentrating only upon those issues and variables judged to be the most significant.

In England, until very recently, the market-orientated patterns of production fostered by the Common Agricultural Policy of the European Community have had little in common with the 'welfare state approach' (Sewell and Parker, 1988, p. 784) that characterized the provision of water services for public health, economic development and urban expansion. The Thatcher government, however, has changed the overriding goals towards which water management works to include greater regard for economic and market principles. This has been achieved through the establishment and ruthless application of a new system of 'ministerial' accountability' (Sewell and Parker, 1988, p. 785) and in the process the institutional structure for water management in England has changed from a model for reorganization elsewhere (Okun, 1977; Sewell and Barr, 1977) to a model of how a hostile government can undermine concepts of holistic resource management.

Integrated water management

Ideas of integration have centred around the needs of the water supply
service through river basin planning and water quality control facilitating
inter basin transfer, whilst other possible avenues of integration have been
neglected. There has been no overall approach to pollution control and
environmental quality until very recently (Royal Commission of Environmental Pollution, 1988). There has been no overall co-ordination of water
resources planning with the needs of regional economic development as in
neighbouring Scotland. Indeed, links between water management and
the planning of urban and suburban development seem to have been
weakened over the last decade or more.

A lack of integration with agriculture

The official approach to agricultural land use is changing dramatically at
the present time. In recent years production grants have been withdrawn
(although guarantee prices remain), to be replaced by farm-diversification
grants to encourage recreation and tourism and other non food rural land
uses, farm-woodland schemes, conservation grant schemes and set-aside
schemes, taking land out of production. Nevertheless, a serious situation
has arisen over previous decades. In 1985 a water quality survey revealed
that the hitherto steady improvement in river quality had faltered, probably for two reasons: unsatisfactory agricultural practices and sewage
treatment works (DOE, 1986a). Furthermore, the Environment Committee of the House of Commons found in 1987 that agricultural pollution,
declining quality of sewage effluents and inadequate monitoring and prevention of pollution on the part of the authorities were the three serious
threats to the continuing quality of rivers (House of Commons, 1987).
    The problem may be illustrated succinctly with respect to an investi-
gation by the South West Water Authority into the decline of salmon
fisheries in the Rivers Tamar and Torridge (ENDS, March 1986, p. 5).
Agricultural intensification was blamed as the main cause of the deterioration. In the area, livestock numbers had doubled between 1952 and 1982,
producing in the Torridge basin a potential waste disposal problem
equivalent to a town of 600,000 people or forty times the population
actually connected to sewage treatment works within the catchment.
Intensive livestock rearing had also brought greatly increased levels of
fertilizer application to pasture and increases in silage making over the
1970s. To cap all this, schemes of improved drainage brought faster runoff.
In such ways inadequate arrangements for nonpoint sources of pollution
have emerged as a major issue, not least with respect to nitrates. A system
of water protection zones is now proposed for implementation (Knight and
Tuckwell, 1988; DOE, 1988a). The idea is to control activities such that

runoff is not so contaminating within specially defined areas of particular importance as sources of water supply.

Agricultural intensification has clearly had a great impact on water management over recent years but such change was not integrated in any structure of policy making. Indeed, on the contrary, Kinnersley (1988, p. 72) has suggested that the relationship between the government ministries responsible for the two spheres, the Department of the Environment and its predecessors and the Ministry of Agriculture, Fisheries and Food has been one of rivalry and contest.

The broad sweep of events

The story of developing institutional arrangements is complex. In this section the main structure of developments is outlined (see Figure 5.1). Demands for increased supplies of water forced a wider scale of operations and planning, and a pattern of water boards gradually grew through the 1950s and 1960s under the auspices of the Water Act 1945 and administrative action on the part of central government. River Boards, established by an Act of 1948, slowly built up a body of hydrological data, and the Rivers (Prevention of Pollution) Acts of 1951 and 1961 brought a system of control over the quality of surface water. The Water Resources Act 1963 added new river basin based functions of water resources planning and river abstraction control to facilitate a national plan for water resources development and inter-regional transfer. To implement these, a new set of institutional arrangements were enacted in 1973. No sooner had these ten new regional water authorities (responsible for the full range of water management functions) established themselves when the growth in demand they had been created to respond to, collapsed. In addition, recession and inflation engendered unprecedented financial pressures.

In 1983 the Thatcher government recast the management of the regional water authorities on business lines. The new breed of water business managers were soon suggesting that privatization would lead to a better deal for consumers, and proposals to transfer the ownership of the assets were made in 1986. The first version was withdrawn prior to the General Election of 1987. The present proposals were announced a few weeks after the victory of the Conservative party and involve a division of agencies into operators and regulators—a return to the form of the original 1963 arrangements, but with significant changes in the scale of operations and operating ethos of the agencies involved. Ten 'Water Services Private Limited Companies' (WSPLCs) will manage clean and dirty water services. A new National Rivers Authority, a national equivalent of the former regionally based river authorities, will take over regulatory and conservation functions.

LEGISLATIVE	ADMINISTRATIVE
The Water Act 1945	
The River Boards Act 1948	
	Voluntary amalgamation of water supply agencies 1950 to 1970
The Rivers (Prevention of Pollution) Act 1951	Controls over NEW point sources (only)
The Water Resources Act 1963	Abstraction Licensing 'Minimum Acceptable Flows' for the dilution of pollution A National Plan for Water Resources
The Water Act 1973	Ten Regional Water Authorities and integrated river basin management
The Control of Pollution Act 1974	
	River Water Quality Objectives 1978
	External finance limits and stipulated rates of return on assets employed 1980/1
The Water Act 1983	Press and public excluded
	Public Registers of pollution control conditions (authorized 1974) made available to the public 1985
	First White Paper on privatization published 1986, withdrawn 1987
The Water Bill 1988	National Rivers Authority Statutory Water Quality Objectives Water Services Companies

Figure 5.1 A chronology of major events since 1945

The integrated approach to assuring future supplies of water

The administration of water supplies lay in the hands of a large number of different types of authority until 1973 when Regional Water Authorities (RWAs) took over the service except in 29 areas of private companies. These escaped the dominant trend of municipal take-over in the nineteenth century and serve significant areas in the prosperous south-east of the country as well as in the north-east. These areas excluded, the typical water undertaking had been the local authority. In many areas, when the time had come in the years of expansion and economic prosperity that characterized the 1950s and 1960s to extend their sources of supply, water undertakings had merged with their neighbours to form joint boards rather than bear the full costs of developing new sources alone. The Water Act 1945 had given the central government ministry responsible for water powers to enforce compulsorily such mergers when it appeared in the interests of optimum resource management. These had been used sparingly, and it was by persuasion and voluntary action that the number of independent undertakings fell from nearly twelve hundred in the late 1940s to fewer than two hundred in the early 1970s.

Water resources planning

By 1960 it was clear that the *ad hoc* and largely localized approach required some alteration if the expected rapidly expanding demand for water was to be met. Not only were standards of living rising and leading to increases in per capita consumption of water but the clearance of slum housing and its replacement with appropriately modern facilities was having a similar effect. Housing and industry were rapidly suburbanizing with consequent shifts in the location of demand. Development of demand was transcending local authority boundaries. The Water Resources Act 1963 strengthened the role of river boards with regard to the control and planning of water resources. The 32 river boards were reformed into a pattern of 29 River Authorities (see Figure 5.2). Powers were given to license the abstraction of water from river basins and the authorities were required to create a water resources development plan for their river basins. The intention was that the River Authorities would eventually act as water wholesalers, and having identified and built optimum resource developments for their area they could then sell portions of bulk supplies of water to constituent water undertakings as required (Craine, 1969).

At the national level, the Water Resources Board was also created with the task of collating the efforts of River Authorities and producing plans at first for major parts of the country, such as the North, or the Midlands, and then, ultimately, for the whole country in a National Plan published in 1973 (Water Resources Board, 1973). It seemed that the demand for water would double in the thirty-five years or so to the end of the century. Fears began to grow as to whether the daunting task of doubling the capacity of developed catchments could be achieved in time, especially in the light of the growing criticism and opposition that proposed developments were facing. Up and down the country schemes were being opposed by a variety of interests (Mitchell, 1971), and such opposition could delay a project by as much as five years.

This, plus the propensity for local representatives of water undertakings to disagree and take what some might call an overly parochial perspective, seemed to suggest that the original concept of the optimum resource development was in jeopardy. A research programme undertaken by the Trent River Authority had shown the potential of water re-use in the

Figure 5.2 The River Authorities, 1963–73

context of integrated river basin management (Trent Steering Committee, 1974). It then emerged that the state of sewage treatment works through-out the country was such that several key elements of the emerging national plan for water (see Figure 5.3) involving the transfer of water over large distances using natural river courses could not proceed without the heavy additional cost of rebuilding works (Ministry of Housing and Local Government, 1970).

Implementing the plan

The national water plan envisaged a single source supplying many auth-orities through elaborate systems of inter-basin transfer and groundwater

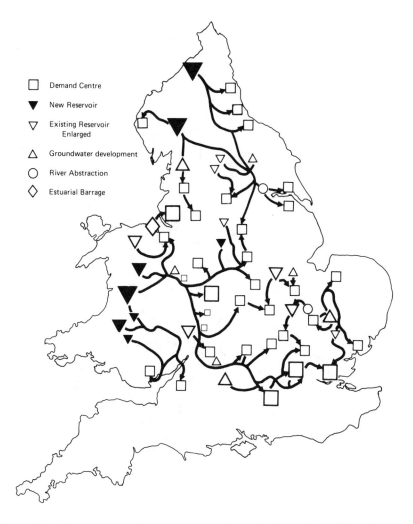

Demand Centre

New Reservoir

Existing Reservoir
Enlarged

Groundwater development

River Abstraction

Estuarial Barrage

Figure 5.3 The principal elements of the National Water Plan 1973. (Source: Water Resources Board, 1973, *The Water Resources of England and Wales*, 2 volumes, HMSO)

storage. Co-operative ventures on this scale had no precedents. The management of water quality in rivers, especially the careful regulation of the performance of sewage treatment works, would require hitherto unheard of degrees of co-ordination and integration of investment and operating plans. With the benefit of hindsight, economist Kinnersley (1988, p. 90) has called the plan 'not a future strategy but an engineering spree'. Rees (1977) also highlighted how expensive such a scheme of construction was going to be and called for more consideration of waste control and demand management strategies. Nevertheless, at the time, the case for change arising from the water plan was very strong. Not only this,

Figure 5.4   The Regional Water Authorities

but the extent to which local government as a whole was adequate to deal with the demands of the second half of the twentieth century was increasingly being questioned (Royal Commission on Local Government in England, 1969). If local government were to change, the question arose as to what should happen to the water supply, sewerage and sewage treatment functions of local authorities. There was a strong argument for including such functions with planning and the other services that amounted to necessary infrastructure for expansion and development, such as roads, schools and public housing. This is what happened in Scotland (Sewell, Coppock and Pitkethly, 1985). In England the water services were seen as appropriate for more radical restructuring.

The Water Act 1973 brought new authorities, as shown in Figure 5.4. For detailed discussion of the establishment of the authorities the reader is referred to Okun (1977), Kromm (1985) and Sewell and Barr (1978).

Faltering demand, a new financial regime and new ideas

The economy of the United Kingdom ran into serious trouble in the late 1970s, notwithstanding the rapid development of North Sea oil resources. Serious rounds of public expenditure cuts began to affect the water industry from 1977 onward. At the same time economic growth gave way to recession. The purpose for which the structure had been created, the rapid expansion of capability to supply water, collapsed as recession deepened and, some would say, a process of rapid de-industrialization accompanied the rigorous policies applied by the Thatcher government. The fall in industrial demand for water and the slowing of growth in the domestic sector is evident in figures for metered water supply. This fell from 5400 Ml/d in 1971 to 4500 Ml/d in 1985. The new Conservative government of 1979 had clear economic priorities. A major plank in the strategy to return the country to economic health was rigorous control of the public sector borrowing requirement, including the amount of capital required for improvements to water, sewerage and sewage treatment. External finance limits were placed on the RWAs which reduced the extent to which developments were financed by long-term borrowing (as opposed to financing directly out of revenue) from 40 per cent in 1981/2 to 10 per cent in 1986/7. Water rates rose as consumers paid for new developments and the replacement of the Victorian infrastructure as required. Furthermore, the Government introduced a system of stipulated rates of return on the net current value of assets from 1981/2. Assets were valued according to their current replacement cost. Rates of return were initially set at or below 1 per cent but have been steadily increased since so that common rates were around 2 per cent in 1986/7.

Under new management

As part of its general approach to nationalized industries, the Government promoted the Water Act 1983. This concerned the composition of the boards of the RWAs. Previously these had been composed of representatives of the local authorities within their regions. Another third or so of the members continued the tradition of the River Authorities and the River Boards before them of bringing together representatives of all the interests involved. These were industry, landowners, farmers, anglers and nature conservationists and the like. The 1983 Act replaced these rather unwieldly and representative boards with much smaller and more businesslike groups of ten to twelve appointees. Chairmen were expected to take on an executive role and a new breed was recruited from industry.

It was not long before these new executives were pointing out that the water industry could provide its services at more reasonable prices if it were set free from government imposed restrictions. Government was inclined to agree and proposals to privatize the ten RWAs were published in 1986. The multi-functional authorities would continue to discharge the same mixture of revenue earning services, regulatory and control functions plus public duties of recreation provision while maintaining conservation interests. There were to be two major sets of safeguards: the quality of river water was to be specially protected and a Director General of Water Services would grant operating licences to the new private companies which would be conditional on quality of service standards being kept and constraints on the level of charges observed (DOE, 1986b).

These proposals satisfied neither those interested in rivers nor business critics who were fearful of the principle of one set of private companies imposing regulatory requirements and were apprehensive as to future levels of charges. The first set of proposals was withdrawn and replaced in July 1987 with the current plan. Ten Water Services Public Limited Companies (WSPLCs) will provide water supply, sewerage and sewage treatment services. Other functions will transfer to a new National Rivers Authority. The Director General of Water Services will license the WSPLCs, supervise public complaints and price increase rather in the same manner as new public officials have recently been appointed to perform similar functions with regard to the recently privatized gas and telephone services.

River basin management

Policing water quality

An effective and comprehensive system for controlling the quality of river water in England did not come until the 1960s and its effectiveness remains controversial. The Rivers (Prevention of Pollution) Act 1951 brought temporary controls over new discharges to surface waters. Since the great majority of industrial effluents and outflows from sewage treatment works already existed, and since the waste waters of new developments were most often directed to existing sewage treatment works (STWs), this had much less impact than might be imagined. The system was that any new discharge would be illegal unless it had the consent of the local River Board. Consent could be conditional on quality standards being met and could be reviewed from time to time. Conditions were subject to appeal to the Secretary of State lest an industrialist or STW operator felt aggrieved. The River Boards had been established in 1948 and had taken over responsibility for pre-existing functions concerning fishery management and land drainage. They had already started the job of collecting the detailed hydrological information that would be required to operate the pollution control system, the need for which and its relationship to future water supplies having been discussed by the CAWC report of 1943 (CAWC, 1943).

The consent system was originally supposed to be temporary. It was supposed to be an interim, holding measure while by-laws, applying to small sections of river separately, were drawn up which would designate prohibited substances and/or standards which had to be met before effluent could be discharged legally. Different sets of by-laws would apply to different parts of the river basin to take hydrological variation of flow available for dilution and its ambient quality into account. The staff required to formulate such an elaborate, permanent system were not made available and little progress had been made by the late 1950s when a series of dry years once again focused attention on the need to integrate the management of quality of river basins with the planning of future supplies of water (CAWC, 1960; CAWC, 1962). The decision was taken to abandon the by-law system and instead extend the consent system to all existing sources of effluent. This had the merit of practical flexibility, was simple to operate and cheap to administer but raised questions of fairness and comparability (Storey, 1979), especially since under some spurious notion of the importance of industrial trade secrets, the exact content of consents and their conditions was not to be made public. The Rivers (Prevention of Pollution) Act 1961 implemented the system but it was several years before consent conditions were devised for all sources of effluent.

The 1963 Act made possible a more rigorous management approach to rivers. River authorities were to use their greatly expanded data collection function to establish 'minimum acceptable flow' levels for each part of their river basin(s) which would act as planning base lines around which resource development and water quality targets could be set. Powers of construction could have been used to control the amount of dilution water available to absorb the effects of pollution downstream, or for other integrated purposes such as flood control, but they were never realized. Such a role was not felt feasible. Under the management of the River Authorities some progress was made. A base line survey of the quality of rivers was undertaken in 1958 and subsequent surveys showed slow but real progress by 1970, again in 1975 and then again by 1980 (see Table 5.1).

The whole system operated, however, under a shroud of secrecy with only an occasional glimpse of how it really worked. One notable view was provided by Tinker (1975). On taking over from its predecessors, the Severn–Trent Water Authority asked for the voluntary co-operation of dischargers to publish the position with regard to river water quality in its area. The result was that only a 'thankful thirteen' (out of 63 agreeing to participate) industrial premises were publicly acknowledged to be com-

Table 5.1 The river water quality surveys

	1958		1970		1975		1980		1980*		1985	
	km	%	km	%	km	%	km	%	km	%	km	%
Unpolluted	24,950	72	28,500	74	28,810	75	28,810	75	28,050	69	27,460	67
Doubtful	5,220	15	6,270	17	6,730	17	7,110	18	8,670	21	9,730	24
Poor	2,270	7	1,940	5	1,770	5	2,000	5	3,260	8	3,560	9
Grossly polluted	2,250	6	1,700	4	1,270	3	810	2	640	2	650	2
Length surveyed	34,690		38,400		38,590		38,740		40,630		41,390	

*amended system
Note: Non-tidal water only—not including canals
Source: Kinnersley, 1988, p. 207

plying with their consent conditions every time a sample had been taken. A further 11 industrial discharges had not yet been given consent conditions (this a dozen years after the relevant legislation). A 'nasty nineteen' sewage treatment works were identified as causing major problems. Not one of these had ever been prosecuted. A 'furtive forty' companies would not allow the disclosure of details of one or more of their effluents. The picture painted is a dismal one of relaxed river authorities doing what they could to improve water quality but refraining from prosecutions when the interests of a local authority (responsible for the STWs) and local industrial enterprises seemed threatened.

It would seem that, like one hundred years before (Law, 1956), the system of consents was not a major burden on those who polluted the rivers of England in the 1960s and 1970s. Nevertheless, there was a permanent, if weak, pressure to produce improvements and a substantial amount of work was done in rebuilding STWs. In the early 1970s it looked as though the effort would be intensified, but financial pressures conspired to produce exactly the opposite effect. The simultaneous rationalization of industry, however, meant that the pressure on various stretches of water and on many STWs was considerably reduced as factories closed down and were demolished. It was not until 1985, therefore, that the effects of stringent control on investment in improving water quality were obvious in the river water quality survey (see Table 5.1). The reorganization of water and sewerage into ten authorities in 1973 raised the obvious conflict of interest between the supervision of river water quality and the management of investment in sewage treatment works (the principal source of pollution). Independent water quality panels were established within the Regional Water Authorities, the idea being that the public could be reassured by this committee of the great, the good and the wise that nothing improper with regard to the maintenance of the quality of river waters was taking place.

Just in case anybody was not entirely convinced by this, there quickly followed the Control of Pollution Act 1974. This was a comprehensive review of powers concerning the environment with sections on air pollution, waste disposal and noise as well as water pollution control. Essentially, the existing system and powers were re-enacted but with two important additions. First, a consent system of control was extended to sewers, making STWs much easier to manage, and in following years a system of charging was also applied. Second, the veil of secrecy over consent conditions was to be lifted. A public register of such consents was to be established so that private individuals or clubs could easily find out what was supposed to be happening and take their own test samples to monitor the performance of the regional water authority. In addition, it became possible for such private groups to mount a private prosecution in the event of their feeling that the water authority was acting unlawfully.

The 1974 Act was to be implemented in stages as and when the Secretary of State for the Environment thought appropriate. The public registers section was not implemented until 1985. Meanwhile, to make the picture suitable for a viewing public, a review of all existing discharge consent conditions was launched in February 1978 (Bates, 1979). Regional Water Authorities (RWAs) were to specify river quality objectives and then reset conditions in the light of these. The quality objectives were to be the baselines for the planning of future investment. The view emerged that an authority could not be prosecuted for failing to operate an STW incorrectly if the water quality objective was being met. It was originally hoped to bring COPA into effect by 1980 but then further cut-backs in investment began to bite and implementation was again postponed. In 1983 the Secretary of State for the Environment took to himself the power of granting consent to discharges from water authority sewage treatment plants. There followed further relaxation of consent conditions in 1983–4 and a view emerged that effluents of STWs would not be considered unsatisfactory unless they failed to meet consent conditions on more than 5 per cent of occasions. Notwithstanding this message of the figures, however, a fifth of STWs are recorded as unsatisfactory (House of Commons, 1987, p. xvi).

Anecdotal evidence suggests that little has changed since the 1970s. The *Yorkshire Post* (8 Nov. 1988, p. 1) reported:

many of Britain's biggest companies are illegally dumping thousands of gallons of dangerous chemicals, pesticides and other highly toxic waste into rivers and streams —and getting away with it . . . The true extent of the disregard for laws designed to clean up polluted waterways is revealed for the first time by an internal file of offenders obtained by the *Yorkshire Post*, which shows that 142 companies and public bodies in Yorkshire, Humberside and North West Derbyshire were in breach of the 1974 Control of Pollution Act in the 12 months to August.

The initial proposal to privatize the water services, and with them the regulatory functions concerning water quality control, was met with some disbelief by those familiar with water management. The Government claimed that river quality would be protected in a manner more effective than ever before. Their proposals that published river quality objectives would be given a statutory basis and rigorously monitored by the Department of the Environment failed to command public confidence. Even friends of the Government, such as the Confederation of British Industry, were not impressed. They saw the new private companies' powers of prosecution for water pollution as unacceptable and wanted to see separate regulatory authorities (ENDS, Aug. 1986, p. 139).

The modified privatization proposals are notable for rigorously separating the regulatory role from that of providing services to consumers. The final abolition of the contradiction of the same authority seeking to keep costs as low as possible whilst at the same time setting itself what are supposed to be the highest standards of environmental quality has focused more public attention on the operation of the pollution control system than at any time over the last forty years. A national scale of operations was appropriate for a number of reasons. Those that are to be regulated, such as agriculturalists and industrialists, are already well and effectively organized at the national scale. The need for consistency in the way that the utilities were treated throughout the country, and the fact that the risks and threats to rivers were not respecters of water authority boundaries, were also important considerations; but overall: 'If we are to take a more rigorous approach to the enforcement of water environment legislation . . . a national body will be better placed to develop water environment policies coherently and implement them effectively' (Belstead, 1988, p. 239). The Water Bill, when published, apparently took some people by surprise as to the strictness of planned regulation. The *Financial Times* of 21 November 1988 quoted one water authority chairman as saying, 'We can just about tolerate the bill as drafted but the scale of regulation is very harsh'. Furthermore, another 'industry leader' is quoted as saying 'The whole bill is drafted as if the PLCs will be run by a bunch of crooks'.

Wider aspects of river basin management

The wider aspects of river basin management—wildlife and conservation, fisheries, navigation and recreation—have typically received little atten- tion. The Nature Conservancy Council told the House of Commons Environment Committee (1987, p. xiii) that in practice: 'Wildlife is con- sidered almost as an add on'. The first privatization proposal dismissed them, for example, in just over a single page (DOE, 1986b). In this review, wildlife and recreational issues receive similar short shrift because they have not figured largely in demands for restructuring or for a greater degree of integrated management. The reorganization of 1973 brought contradictory signals with regard to recreational uses. On the one hand the new regional water authorities were given a statutory duty to pay due regard to the needs of recreationists and those interested in wildlife at the same time as the main arena of aquatic leisure was kept out of the new framework. Canals were retained as the responsibility of the British Waterways Boards (with which they remain) despite the extent to which their use had become nearly wholly recreational and as sources of indus- trial water supply. Fears that the national network of routeways would be broken up overrode the benefits of integrating them into the drainage and water supply patterns of their regions, not to mention any concept of holistic planning of recreational opportunities.

Since 1973 many people have begun to make the connection between a day out in the countryside and the use of water authority land, and by 1988 something of the order of 100,000 hectares had been made accessible for outdoor pursuits (*Independent*, 1 Dec. 1988). The potential loss of access or of commercial development or, worse, the possibility of it being turned over to paying customers only, has proved an early focus of opposition to water privatization from the outdoor pursuits lobby. The Council for the Preservation of Rural England commissioned research which suggests that there will be pressure for high value recreation facilities, rather than the continuance of good conservation and free access. It seems that the public may look forward to power boating and time-share log cabins (Bowers, 1988).

The dirty water disposal and treatment services

Local authorities had been building sewage treatment works steadily over the years but too many local authorities were too small to be able to afford to invest in proper facilities and too many could not afford the services of fully trained professional specialists to run the systems that they already had. The report of a working party on sewage disposal appointed in 1969 was aptly entitled *Taken for Granted* (Ministry of Housing and Local Government, 1970). It was estimated that some 3,000 sewage treatment works were failing to meet basic standards (p. 9) and this represented 60 per cent of the total. The principal cause of failure was simple overloading and failure to renew infrastructure appropriately.

As has been explained earlier, the radical restructuring of management and resources available to the sewerage and sewage treatment services was a principal objective of the 1973 reorganization. But programmes of refurbishing sewers and rebuilding sewage treatment works had barely begun when the financial crises of the late 1970s struck. Expenditure on investment in sewage treatment fell quite dramatically in the late 1970s to rise again only slowly in recent years. Synnott (1986) has shown how real expenditure on the water services deteriorated by over half from £1,223 million in 1973/4 to £598 million in 1980/1 before beginning to slowly rise again (1982/3 prices). He also traces how the number of sewer collapses around the country increased greatly over the same period. Pearce (1982) has provided a graphic catalogue of crumbling sewers, overloaded and outdated facilities.

Privatization proposals have focused more attention on the dirty services than ever before. Financial analysts fear that the backlog of work required may turn out to be the Achilles' heel of the whole deal, especially if water quality standards set by the new National Rivers Authority are anything less than a farce. ENDS (Sept. 1987, p. 12) say 'The Government's drive to get costs down has turned into a loaded gun pointing at its own foot in the context of privatisation'. In a clever presentation ploy, government spokesmen began to claim that privatization will help environmental goals because their achievement will no longer be linked to constraints on public expenditure. Nevertheless, a good deal of accounting of what remains to be done continues. Early results have been alarming: about 1,000 sewage treatment works require upgrading to meet consent conditions and a programme of improvements to sea outfalls is required to bring beaches up to standards set by the European community (ENDS, April 1988, p. 3). Add to this the expenditure required to bring water supplies up to required European standards and one begins to see why the chairman of the Water Authorities Association has estimated that charges to consumers will need to rise by at least 50 per cent in real terms to give the authorities a chance of meeting the standards.

In anticipation of privatization and to allay fears of the effect of the sewage issue on the success of the privatization exercise, the Secretary of State for the Environment announced a £1 billion programme for improving the performance of sewage treatment works, over the period to 1992. This is in addition to the £2.2 million already authorized for 1985–9. In the meantime, the Secretary of State feels there should be no prosecutions until the water authorities have had time to put their houses in order. It would seem that more dramatic improvements in updating STWs is about to be made in the last seven years of the existence of RWAs than was made over the previous twelve. Privatization may offer the key to future progress. Many existing STWs and other water authority installations are on land which has significant potential for commercial or residential redevelopment, especially in London and the Home Counties and in the high amenity national park or green-belt areas. The sale of such assets might finance long overdue schemes of regional sewage treatment and disposal.

Factors in water management

Water Authorities themselves seem to have been caught rather off balance by the fluidity of privatization proposals. At first they were largely in favour of keeping all functions within the new private WSPLCs because of the benefits of 'integration' that this would bring. Thames RWA ran a campaign of corporate advertising highlighting its integrated approach and inviting the public to become familiar with one of their most demanding customers—a salmon returning to the Thames—something that had been made possible apparently only through integrated river basin management. Then they gradually changed to the view that a separate national regulatory body would be in everybody's best interest, notwithstanding the loss of certain economies of scale following the loss of laboratories, technicians and field officers to the new regulatory authority.

The Severn-Trent RWA told the House of Commons Environment Committee (1987, p. 226) that the authority was convinced that the present integration of catchment regulation of water quantity and quality with the operational functions of public water supply and sewage treatment is both

efficient and beneficial. The benefits of such integrated catchment management are practical, they said, and cannot readily be separated from the
operational side of the Authorities and their financing. This seems to mean
that the benefits relate to the sharing of scientific staff, laboratories,
vehicles and the like. Similarly, the Yorkshire Water Authority also
proclaimed its commitment to integrated river basin management (House
of Commons, 1987, p. 321):

> By looking at a catchment as a whole, the interaction between discharges from
> STWs, storm sewage overflows and trade premises, along with abstractions from
> rivers and the release of compensation water from impounding waters can be
> studied and the most effective means can be found of achieving and protecting the
> quantity and quality needs of river users . . . Integrated river basin management
> provides a framework for the control of abstractions and discharges so that water
> quality objectives are achieved and maintained according to the 'Best Practicable
> Environmental Option'. . . For effective quality control, standards should be based
> on the best scientific information and be realistic for practical river management.
>   (House of Commons, 1987, p. 321: reference to 'Best
>   Practicable Environmental Option' made in the context
>   of Royal Commission on Environmental Pollution, 1988)

Soon after, water authorities were supporting proposals to break up the
integrated authorities in line with Government policy. More recently the
authorities are saying nothing about anything to anybody lest the flotation
of their assets on the stock market be jeopardized. It would appear that
in recent years there has been little opportunity to put principles of
integration into practice except in the sense of sharing resources, often of a
mundane type such as laboratory equipment, vehicles and offices. A clear
example of an integrated strategy in operation is difficult to find.

In recent years the English have actually exhibited very little interest in
the concept of integrated water management. Nevertheless, the English
experience may be said to be of value to the understanding of the prospects
for comprehensive and integrated management. In the same way that it
was possible for Sewell and Parker (1988) to evaluate the prescriptive view
of Craine (1969) as to the extent the English were a model for world
progress in institutional design for water management, it is useful to
examine the experience recounted above in terms of the framework
provided by Mitchell (1987).

The scope of comprehensive and integrated water and land management

Clearly, land and water resources are better managed in an holistic manner.
The difficulty is that it is generally recognized that any bureaucracy set up
to handle the orderly co-ordination and planning of all relevant inputs
tends to find its task humanly impossible (March and Simon, 1958; Braybrooke and Lindblom, 1963; Faludi, 1973): the 'rational-comprehensive

ideal' of planning and policy making is extremely difficult to achieve and more likely to end up as a 'satisficing' compromise resting on factors far removed from the 'facts' of the situation. Central planning is a well documented failure world-wide, not least with regard to multi-functional and large-scale water resource development. It is much better to focus on more limited integration to achieve priority objectives. The logic of the 1973 structure established in England was of this nature: integrating river, sewer and water conservation functions to assure future supplies of water. It worked and the achieved integration minimized the effect of drought in both 1976 and 1984. However, in the absence of the expected increase in demand for water, there was no need to maintain water quality objectives as the linchpin of the system of integrated management. A philistine government could force through deeply damaging investment levels behind a curtain of public indifference and official secrecy.

In 1989, water quality was in the frontline of an unprecedented public debate on English water. An avalanche of adverse publicity concerning water quality and the extent to which European standards were not being met engulfed the Government's programme of 'privatization'. Faced with such a deluge at the very time that Mrs Thatcher considers it prudent to lay claim to the 'green' vote and at which the idea of European Integration is engendering serious discussion within the United Kingdom for the first time, the Government have had no choice but to do two things: the old, multipurpose, integrated authorities have been roundly condemned as the cause of a decade or more of institutional inertia (what can you expect from such bureaucracies, ask Conservatives?) and new commitments are being made to a firm programme of 'cleaning up' once and for all (at the consumer's expense, of course, and so as not to be unfair to the new private companies after at least a year of legal immunity from prosecution). Quite accidentally, the Right Honourable Nicholas Ridley, prophet of deregulation and market forces, seems set to go down in history as the Secretary of State for the Environment in England and Wales to do most for the improvement in the quality of surface water.

Boundary problems

The purpose of the 1973 institution of 'river basin management' through the ten regional water authorities was to facilitate interbasin transfer and the integration discussed above. At the same time the river/regional scale in England is sufficiently small such that contact with local land use planning authorities need not have been lost. Recession destroyed the logic of the water resources plan while central government inspired constraints and exhortations destroyed any logic in regional water quality or contact with local land use planning authorities. Accordingly, the new structure reflects the scale of the threat: the water quality control agency will henceforth be a national body albeit operating in regional divisions (initially at least corresponding to the present regional authority boundaries—themselves hydrologically defined). This leaves the spatial pattern of the new private water services companies desperately short of logic except in the Thames area.

In the latter, many private companies serve important commuting areas for London, leaving the new company to serve the interstices between them and the capital itself. This looks like a healthy pattern of several adjacent companies competing with each other to report the most favourable balance sheets and quality of service indices. Elsewhere in the country the monolithic regional authorities (now renamed, in Thatcher newspeak, 'water businesses of England and Wales') have no logic. They have no identity with cities and regions, nor with county towns and their surrounding semi-rural dwellers: they have no community of interest and they have no identity in the public mind. This may be why the enormous effort through television and lavish press advertising campaigns has badly backfired, with the commercials concerned said to be the most unpopular ever and, indeed, the whole issue of water 'privatization' said to be a major factor in the sharp decline in the opinion poll ratings experienced by the Government in 1989.

Institutional arrangements

There has been great faith in a kind of 'structural determinism' in English water management. The solution to successive problems has been seen in institutional change. Corporate planning and public interest managerialism came and went in the 1970s to be replaced by commercial imperatives and regulatory bodies. It seems possible that history is about to repeat itself with a structure for water management being established with an eye to yesterday's problems rather than tomorrow's. No system can survive a hostile environment and sustained attack from either public or politicians. The latter say they are about to take a backseat on water management and it remains to be seen what happens to the industry now public anger has been aroused. The Government expects that interest will dwindle with time as it seems to have with regard to British Telecom, the privatization of the nation's telephones having aroused similar passion at a similar stage in the process of its privatization.

Organizational structure and participant attitudes

As Mitchell has pointed out (1987, p. 9), factors such as the context in which institutional arrangements have to operate, the mechanisms applied by management to problems, the principal components of organizational culture in different parts of the system and which wings are in the ascendancy and which in decline and the balance of sources of legitimization that is sought, from an array of professionalism, popular public approval, political priority and value for money, are all important elements shaping the attitudes and perceptions that determine whether an institutional system works and survives. Recent English experience shows such factors at work in not one but three historically distinct phases: (1) with regard to the creation of the integrated system of ten regional water authorities, (what might be called the age of technical supremacy); (2) the survival of these authorities after their original *raison d'être* had disappeared, (what might be called the age of the accountant); and (3) the current set of proposals to change to the discipline of market forces where appropriate and consistent with adequate environmental and consumer protection through rigorous regulation (reorientation to market principles). The English story is one of wider social and economic contextual issues producing radically changing demands on the system, forcing changes in structure and mechanisms which, in turn, requires change in the culture of the organizations involved. That change then demands an appeal to new contextual factors for its legitimation.

The age of technical supremacy: 1973–78. The approach was one of managerialism replacing notions of local public service and an ethic of value for money replacing one of a primary focus on protecting the public health. The mechanisms of the new regional water authorities seem to have been characterized by several tendencies: a tendency to internalize externalities perceived to be important, hence river conservation was included but agricultural land use was not. Flood control and land drainage were in but town and country planning, the location and servicing of urban expansion and the co-ordination of efforts to maintain and enhance environmental quality were out. Second, there was a tendency to depoliticize potential conflicts by getting everybody with an important interest represented around the table of a committee or subcommittee. Hence we had locally elected representatives on the main boards of the authority representing the ratepayers and the former owners of the assets. We also had the Secretary of State appointing representatives of industry, agriculture, landed estates and so on. We had the statutory enshrinement of fisheries and land drainage subcommittees to ensure influence on the part of the interested parties. We did not have any open democracy. Third, particularly with regard to river quality, there was a tendency to rely on processes of consultation, bargaining and compromise; a tendency to operate in an envelope of confidentiality—secrecy to those not in the inner circle; a tendency to avoid open disputes and avoid resort to the courts; and a tendency to require and apply standards that were based on the capabilities of existing technology and knowledge rather than standards that would force innovation. This is the administrative style of the urbane professional. The 1973 reorganization was largely a process of getting rid of irritating parochialism on the part of local councillors and councils and replacing it with a technical meritocracy.

The early culture of the regional water authorities was one of smug self satisfaction: a 'just leave the job to us chaps, we know what we are about' approach. Hence, as late as 1982, the Association of Water Officers in conference could be told by the Chairman of the National Water Council, Sir Robert Marshall, 'We are almost too much for our own good the silent inconspicuous service'. Having deliberately eschewed democracy and elected representatives, the RWAs had an increasing difficulty in legitimizing their actions in the late 1970s. The initial rationale of technical competence had been in good company. In local government, the Town and Country planning system was simultaneously reorganized around the concept of 'structure plans', strategic policy documents setting the framework of future developments over a wide range of policy matters from transport through housing to social services. Local government as well as the water services had been reorganized along lines based on principles of management teams, operating a rolling programme of objective-orientated initiatives within the context of an overall strategic plan. The difference in the water industry was that when things started to go wrong with the plans, when the context began to change dramatically so that they were no longer relevant, local government could go back to an appeal to local democracy for its justification. The water authorities were left in a vacuum.

The age of the accountant: 1979–89. Economic recession and the collapse of industrial demand for water destroyed the strategic rationale of regional water authorities. Severe pressure on levels of public expenditure from 1977 onwards also increasingly undermined and challenged the possibility of exercising technical supremacy in other aspects of the authorities' work, notably sewerage, sewage treatment and the enhancement of river quality. Although a new structure did not come until the Water Act 1983, the water industry in England began to come under new management from 1979 onwards. The Thatcher government believed strongly in reducing the role of state expenditure in the economy as a whole, in ensuring 'value for money' as measured by standard accounting means in those areas of state endeavour that remain, in a programme of disentangling government from much of the strategic direction of economic activity as soon as possible, and in imposing the discipline of customer choice or market forces to the greatest extent possible in as many areas of public service as possible. All of these are major contextual changes with profound implications for the operation of water authorities in the 1980s. As soon as the technical meritocracy had established itself it became doomed by chill new demands.

Developing a style for the 1980s was a matter of gradually overlaying a new set of procedures on the old tendencies and letting go a lot of staff of all grades and types. (Total water authority staff fell from 63,000 in 1979 to less than 49,000 in 1988.) Making the systems of stipulated rates of return on capital assets deployed and external financial limits work involved rigorous attention to cost control, staffing levels and the justifiability of investment and maintenance practices and programmes from the perspective of economic productivity and prudence, not necessarily technical efficiency. The new breed of water authority chairman brought in after 1983 by the Secretary of State seems to have seen his job as sorting out the industry. Details and instances illustrative of the new approach are rare, since one of the first measures the new men took was to abandon public access to water authority meetings and close them to the press. The changing context brought new budgeting procedures and the rationalization of procedures, staffing and costs generally. The new men have been spectacularly successful in raising the rates of profit recorded by their authorities.

As might be imagined, the new orthodoxies put considerable pressure on the prevailing culture of the water-related organizations, both nationally and regionally. It will be recalled that it was the water authorities themselves that suggested privatization as a way out of the current context. It would seem to have been the Department of the Environment, however, that has endured the greatest trauma. Quite apart from the unenviable task of defending the indefensible over the implementation of the Control of Pollution Act 1974, and the effect of spending on river quality, it may be argued that the administrative culture of the ministry has been quite inappropriate for the objectives the Government has asked it to achieve. Wilks (1987) has provided some useful observations in the context of a study of urban policy. It may be that the city-related divisions of DOE are different from those concerned with water but the argument is interesting, if speculative.

DOE has historically dealt with local authorities, a wide variety of professional departments within them and with other public bodies such as the regional water authorities. Its clients are administrators sitting in working parties or committees. They are not entrepreneurs, hence there is not the same risk of agency capture on behalf of interest groups as is said to have developed between farmers and the Ministry of Agriculture, defence equipment manufacturers and the Ministry of Defence and energy producers and the Department of Energy (Holbeche, 1986; Dawkins, 1987). The function of the department is also relatively straightforward having been deliberately created at the beginning of the 1970s as a 'super-department' to deal with all government involvement with the physical environment. The Department is itself an example of managerialism popular at the time. Its staff are many and varied, and dispersed into many regional offices and different directorates at headquarters. Many small groups of professional specialists exist as islands of expertise in a rich advisory ocean. It is quite impossible for these people to have personal knowledge of the four hundred or so client organizations and therefore, in marked contrast to the same sort of professional advisers in Scotland (Page, 1978), the staff may be characterized as, above all, remote.

Two traditions are said to dominate the style of the organization: decades of quasi-judicial pronouncement (deciding appeals) and decades of cultivating a stance of political insulation or neutrality so as to preserve good relations between central and local government bodies of potentially different political colours. Both of these proved highly inappropriate for the Thatcher years. The present Government's monetary policies required strong central direction on what could be cut and what should be retained. DOE seem to have been unable to provide this.

The Government has been determined to see the burden of high local taxes lifted and in the process has all but dismantled the independence and freedom of local authorities to choose their own levels of service. This, too, required strong executive action from the centre. The Government has been determined not to tolerate left-wing ideas of a 'local state' in which local councils attempt to subvert national policies by going it alone locally in such matters as transport or housing policies. Seven councils, including that for Greater London, have been abolished. Stringent cash limits have been set, legal requirements to put the management of services out to private contract have been imposed, and an array of new powers have been enacted, allowing local people to opt out of local authority services in the areas of housing and education. In short, central–local relations have been transformed. They are nowhere near a state of neutrality now.

A reorientation to market principles in the 1990s.   Some commentators have seen recent events in English water management as a capitalist conspiracy in which concepts of local democracy have been deliberately undermined by regionalization and an appeal to 'criteria of competence': this was then followed by a 'commoditization' of water (Patterson, 1987). It is true that water has increasingly been seen as something which should be bought rather than something which should be provided, but there seems little in the way of evidence for a conscious drive towards market objectives on the part of the industry's participants over the last thirty years. On the contrary, Kinnersley describes the privatization proposals as the result more of a process of stumbling than striving (Kinnersley, 1988). The contextual change that has brought forward the proposals has been entirely ideological. There are no new factors of demand or supply of relevance. The new factor is the debate over how best to manage services. A similar debate is going on over a wide range of public services in England such as leisure centres (Foxall, 1983).

The new market-orientated approach involves a concern with two factors: external motivation (customer or market signals) and internal facilitation (organized market research and the careful planning of achievable objectives). Such criteria seem entirely consistent with the proposals for Water Services Public Limited Companies providing clean and dirty water services. But one is only likely to find a truly market-orientated approach if the organization depends on it for its survival and there must be some question whether the new Director General will ever dismiss a licensed company no matter what its performance.

Such criteria are at variance with the requirements of regulation in the interests of environmental quality or conservation. The latter pursuits must, however, equally ensure that they avoid the classic deficiencies of public service management, such as a fixation with internally set objectives and criteria, a preponderance of vague overall policy goals rather than specific objectives, an absence of thorough research and an absence of a segmented approach to product pricing and availability. All of these seem to characterize the regional water authorities mechanisms of working in recent years.

Prospects for the future. The water services may see a return to ideas of integration if industrial demands for water revive and the National Rivers Authority develops comprehensive plans for river basins or groups of them. With regard to agriculture, it would seem that the days of high subsidy leading to overproduction are over. The new mechanisms of water protection zones, subsidies for pursuing traditional practices in especially sensitive landscape zones and the requirements for impact assessment of major changes (Wood, 1988) could amount to an introduction of a planning system of control over land use if applied with a will to make it so. On the other hand, planning is a concept under sustained attack by the present government. A series of developments have constrained its structure, mechanisms and processes (Blowers, 1987; Thornley, 1988). The balance of development control is said to have swung markedly in the favour of developers and against environmentalists. The system of structure plans providing a regional forum for the examination of key issues such as the maintenance of open country between built-up areas is to go. The once extensive system of regional economic aid to facilitate a policy of equal opportunity has been dismantled (Johnson, 1988).

Conclusion

The important thing to remember about water management in England is that it has never been seen as something of top priority and has only hit the headlines very recently, other than in times of threatened drought. In a system of policy making that depends on items fighting their way to the top of an agenda before they receive serious attention (Solesbury, 1976; Stringer and Richardson, 1980), aspects of water management have rarely managed to overshadow competing concerns.

English water resource management has been an essay in 'muddling through' on the part of smug professionals and not quite the model of rational and comprehensive integrated management that it is often portrayed. Important lessons may be drawn from the English experience on how the rational comprehensive ideal may crumble and be corrupted. The utility of Mitchell's analytical framework can be confirmed. A key element in the unfolding of events has been the professional cultures of engineers and accountants rather than pressure groups, the public or politicians. Generally speaking, water resource management exhibits the key characteristics of the élitist political culture of the United Kingdom: the professionals of the industry have taken advantage of both their low political profile and the official confidentiality that surrounded their every move. That removed the possibility of informed and critical discussion of water policy on the part of the public, thus water managers lost any claim to support when they found themselves challenged by an alien ideology and hence the demise of integrated water management.

References

Bates, G. M., 1979, 'River water quality: maintaining the status quo?', *Journal of Planning and Environmental Law*, 7, pp. 152–8.

Belstead, Lord, 1988, 'Priorities in water and environmental management', *Journal of the Institute of Water and Environmental Management*, 2, pp. 237–9.

Bentil, J. K., 1984, 'Ensuring control of pollution of aspects of the European Community's aquatic environment', *Journal of Planning and Environmental Law*, pp. 707–21.

Blowers, A., 1987, 'Transition or transformation?—environmental policy under Thatcher', *Public Administration*, **55**, pp. 277–94.

Bowers, J. K., 1988, *Liquid Assets*, Council for the Protection of Rural England, London.

Bowers, J. K. and Cheshire, P., 1983, *Agriculture, the Countryside and Land Use: An Economic Critique*, Methuen, London.

Braybrooke, D. and Lindblom, C. E., 1963, *A Strategy of Decision*, Macmillan.

Central Advisory Water Committee (CAWC), 1943, *Third (Final) Report: River Boards*, Cmd. 6465, HMSO, London.

Central Advisory Water Committee, 1960, *Final Report of the Trade Effluents Sub-Committee*, HMSO, London.

Central Advisory Water Committee, 1962, *Final Report of the Subcommittee on the Growing Demand for Water*, HMSO, London.

Central Statistical Office, 1988, *Regional Trends 22 (1987)*, HMSO, London.

Craine, L. E., 1969, *Water Management Innovations in England and Wales*, Johns Hopkins Press, Baltimore.

Dawkins, L. A., 1987, 'The politics of energy conservation and industrial interests in Britain', *Parliamentary Affairs*, **51**, pp. 250–64.

Department of the Environment (DOE), 1986a, *River Quality in England and Wales, 1985*, HMSO, London.

Department of the Environment, 1986b, *Privatisation of the Water Authorities in England and Wales*, Cmnd. 9734, HMSO, London.

Department of the Environment, 1987a, *The National Rivers Authority: The Government's Policy for a Public Regulatory Body in a Privatised Water Industry*, HMSO, London.

Department of the Environment, 1987b, *Digest of Environmental Protection and Water Statistics, 1986*, HMSO, London.

Department of the Environment, 1988a, *The Nitrate Issue, A Study of the Economic and Other Consequences of Various Local Options for Limiting Nitrate Concentrations in Drinking Water*, HMSO, London.

Department of the Environment, 1986b, *Draft Circular from DOE and the Welsh Office—Environmental Assessment: Implementation of EC Directive*, Department of the Environment, London.

ENDS, Environmental Data Services Newsletter, published monthly from London.

Faludi, A. (ed.), 1973, *A Reader in Planning Theory*, Pergamon Press, Oxford.

Farquar, J. T., 1983, 'The policies of the European Community towards the environment: the "Dangerous Substances" directive', *Journal of Planning and Environmental Law*, **11**, pp. 145–55.

Fernie, J. and Pitkethly, A. S., 1985, *Resources: Environment and Policy*, Harper & Row, London, Chapter 8, pp. 140–9.

Foster, S. E., 1985, 'Inland waterways: legal aspects of public access and public enjoyment', *Journal of Planning and Environmental Law*, **13**, pp. 440–60.

Foxall, G., 1983, 'Can the public sector provision of leisure services be customer-orientated?, *The Service Industries Journal*, **3**, pp. 279–95.

Gray, C., 1982, 'The regional water authorities' in B. W. Hogwood and M. Keating (eds.), *Regional Government in England*, Clarendon Press, Oxford, pp. 143–67.

Guruswamy, L. D. and Tromans, S. R., 1986, 'Towards an integrated approach to pollution control', *Journal of Planning and Environmental Law*, **14**, pp. 643–56.

Haigh, N., 1984, *EEC environmental policy & Britain*, Environmental Data Services Ltd.

Haigh, N., 1986, 'Devolved responsibility and centralisation: effects of EEC environmental policy', *Journal of the Royal Institute of Public Administration*, **44**, pp. 197–207.

Holbeche, B., 1986, 'Policy and influence: MAFF and the NFU', *Public Policy and Administration*, **1**, pp. 40–7.

Holdgate, M., 1987, 'The reality of environmental policy', *Journal of the Royal Society of Arts*, **135**, pp. 310–27.

House of Commons Committee on the Environment, 1987, *The Pollution of Rivers and Estuaries*, Third Report, HMSO, London.

Johnston, B., 1988, 'Regional grants flattened by new enterprise wave at national level', *Planning*, 22 January, pp. 10–11.

Jordan, A. G., Richardson, J. J. and Kimber, R. H., 1977, 'The origins of the Water Act of 1973', *Public Administration*, **45**, pp. 317–34.

Keating, M. and Rhodes, M., 1981, 'Politics or technocracy?: the regional water authorities', *Political Quarterly*, **52**, pp. 487–90.

Kinnersley, D., 1988, *Troubled Water: Rivers, Politics and Pollution*, Shipman, London.

Knight, M. S. and Tuckwell, S. B., 1988, 'Controlling nitrate leaching in water supply catchments', *Journal of the Institute of Water and Environmental Management*, **2**, pp. 248–52.

Kromm, D. E., 1985, 'Regional water management: an assessment of institutions in England and Wales', *Professional Geographer*, **37**, pp. 183–91.

Law, B. R., 1956, *River Conservation in England with Special Reference to the Yorkshire Calder*, M.Sc. Thesis, University of London.

Lowe, P. and Morrison, D., 1984, 'Bad news or good news: environmental politics and the mass media', *The Sociological Review*, **32**, pp. 75–90.

Macrory, R., 1988, 'Environmental policy in Britain—the challenge ahead' in DocTer Institute for Environmental Studies, *European Environmental Yearbook 1987*, DocTer International UK, London, pp. 670–2.

March, J. G. and Simon, H. A., 1958, *Organisations*, John Wiley.

Marshall, Sir Robert, 1982, 'Water: the silent service', *Water Bulletin*, April, p. 12.

Miller, C. and Wood, C., 1983, *Planning and Pollution*, Clarendon Press, London.

Miller, D. G., 1988, 'Environmental standards and their implications', *Journal of the Institute of Water and Environmental Management*, **2**, pp. 60–6.

Ministry of Agriculture, Fisheries and Food, 1951, *Land Drainage in England and Wales*, Report of the Land Drainage Legislation Sub-Committee of the Central Advisory Water Committee, HMSO, London.

Ministry of Housing and Local Government, 1970, *Taken for Granted: Report of the Working Party on Sewage Disposal*, HMSO, London.

Mitchell, B., 1971, *Water in England and Wales: Supply, Transfer and Management*, University of Liverpool Department of Geography Research Paper no. 9.

Mitchell, B., 1975, 'An investigation of research barriers associated with institutional arrangements in water management' in B. Mitchell (ed.), *Institutional Arrangements for Water Management: Canadian Experiences*, Department of Geography Publication Series no. 5, University of Waterloo, Ontario, pp. 247–85.

Mitchell, B., 1987, *A Comprehensive-Integrated Approach for Water and Land Management*, University of New England, Centre for Water Policy Research Occasional Paper no. One, Armidale, New South Wales, Australia.

Okun, D., 1977, *Regionalisation of Water Management*, Applied Science Publishers, London.

Page, E., 1978, 'Why should central–local relations in Scotland be different to those in England?', *Public Administration Review*, **12**, pp. 51–72.

Parker, D. and Penning-Rowsell, E., 1980, *Water Planning in Britain*, George Allen and Unwin, London.

Patterson, A., 1987, *Water and the State*, University of Reading Geographical Papers, Number 98.

Pearce, F., 1982, *Water-Shed: The Water Crisis in Britain*, Junction Books, London.

Pearce, F., 1984, 'The great drain robbery', *New Scientist*, 15 March, pp. 10–14.

Penning-Rowsell, E. and Parker, D., 1983, 'The changing economic and political character of water planning in Britain' in T. O'Riordan and R. K. Turner (eds.), *Progress in Resource Management and Environmental Planning, Volume Four*, Wiley, Chichester, pp. 169–99.

Rees, J., 1977, 'Rethinking our approach to water supply provision', *Geography*, **62**, pp. 232–45.

Rees, J., 1985, *Natural Resources: Allocation, Economics and Policy*, Methuen, London.

Richardson, J. J., Jordan, A. G. and Kimber, R. H., 1978, 'Lobbying, administrative reform and policy styles: the case of land drainage', *Political Studies*, **25**, pp. 47–64.

Royal Commission on Local Government in England, 1966–9, 1969, *Report*, Cmnd. 4039, HMSO, London.

Royal Commission on Environmental Pollution, 1979, *Agriculture and Pollution*, Seventh Report, Cmnd. 7644, HMSO, London.

Royal Commission on Environmental Pollution, 1988, *Twelfth Report: Best Practicable Environmental Option*, Cmnd. 310, HMSO, London.

Sewell, W. R. D. and Barr, L., 1977, 'Evolution in the British institutional framework for water management', *Natural Resources Journal*, **17**, p. 395.

Sewell, W. R. D. and Barr, L., 1978, 'Water administration in England and Wales. Impacts of reorganisation', *Water Resources Bulletin*, **14**, pp. 337–48.

Sewell, W. R. D., Coppock, J. T. and Pitkethly, A. S., 1985, *Institutional Innovation in Water Management: The Scottish Experience*, Geo-Books, Norwich.

Sewell, W. R. D. and Parker, D., 1988, 'Evolving water institutions in England and Wales: an assessment of two decades of experience', *Natural Resources Journal*, **28**, pp. 751–85.

Solesbury, W., 1976, 'The environmental agenda', *Public Administration*, pp. 379–97.

Storey, D., 1979, 'An economic appraisal of the legal and administrative aspects of water pollution control in England and Wales, 1970–1974' in T. O'Riordan and R. C. d'Arge, *Progress in Resource Management and Environmental Planning*, Vol. One, Wiley, Chichester, pp. 259–80.

Stringer, J. K. and Richardson, J. J., 1980, 'Managing the political agenda: problem definition and policy making in Britain', *Parliamentary Affairs*, **44**, pp. 23–39.

Synnott, M., 1986, 'Change and conflict in the local planning/water authority relationship since 1979', *Planning Practice and Research*, pp. 14–18.

Thornley, A., 1988, 'Planning in a cool climate: the effects of Thatcherism', *The Planner*, **44**, pp. 17–19.

Tinker, J., 1975, 'The Midlands Dirty Dozen', *New Scientist*, 6 March, pp. 551–4.

Trent Steering Committee, 1972, *The Trent Research Programme* (11 Volumes), Water Resource Board, HMSO, London.

Water Authorities Association, 1987, *Water Pollution from Farm Waste, 1986*, Ministry of Agriculture, Fisheries and Food Agricultural Development Advisory Service, London.

Water Authorities Association, 1988, *Waterfacts, '88*, Water Authorities Association, London.

Water Resources Board, 1973, *The Water Resources of England and Wales* (2 Volumes), HMSO, London.

Wilks, S., 1987, 'Administrative culture and policy making in the Department of the Environment', *Public Policy and Administration*, **2**, pp. 25–41.

Wood, C., 1986, 'Local authority controls over pollution', *Policy and Politics*, **14**, pp. 107–23.

Wood, C., 1988, 'EIA and BPEO: acronyms for good environmental planning?', *Journal of Planning and Environmental Law*, **16**, pp. 310–21.

CHAPTER 6

Institutional arrangements for Japanese water resources

Yoshihiko Shirai

Introduction

The National Land Agency produced the National Comprehensive Water Resources Plan—Water Plan 2000—in October 1987 (National Land Agency, 1987). The plan was developed by the government to clarify the basic direction of future Japanese water resource management in the context of the Fourth National Comprehensive Development Plan.

The Japanese government dealt with the comprehensive control of water resources for the first time in the National Comprehensive Water Resources Plan. The plan advocated the comprehensive control of water resources and the use of the catchment basin as the basic management unit. The rationale is that surface water and groundwater, water quantity and water quality, and water and related land use are all interdependent and should be considered as a total system. The plan also recognized that it is not adequate to focus entirely upon individual catchments in isolation as inter-regional transfers from water surplus areas to water deficit areas are significant (Okamoto, 1983; Shirai, 1987). As a result, it is necessary that the management system be able to consider the possible interdependence among river basins resulting from transfers of water from basin to basin. Moreover, for the establishment of a management system, engineering and behavioural strategies must be considered and used together (Easter, Dixon and Hufschmidt, 1986).

Although a comprehensive approach has been identified in the national water plan, Japan has lacked strong, centralized legislation and organizations to implement such an approach. In that regard, Japan has been different from countries such as England and Wales in which multi-functional river basin management agencies have been established Craine, 1969; Porter, 1978).

In this chapter, the development and control of water resources in Japan since 1945 is presented, with particular emphasis upon institutional and legal arrangements. Over time, laws have been introduced to guide the

148

development and management of water. Prior to the Second World War, agricultural use of water for paddy-fields was given first priority. At the same time, however, hydroelectric power generation, domestic water supply and industrial water supply also received attention. A remarkable increase in water demand after the Second World War led to new legislation to govern water use. Various laws concerning different water uses were introduced, followed by legislation regarding water pollution prevention. These various laws made the realization of a comprehensive approach difficult to achieve. To illustrate the manner in which these different laws directed water management, two examples are examined. The first is the Kitakami River dams which demonstrate a structural management strategy. The second is watershed management in the Yahagi River which is used to demonstrate a non-structural management strategy. Before presenting these examples, however, some general background information about land and water use in Japan is outlined.

Land and water use in Japan

The four major islands of Japan (Hokkaido, Honshu, Shikoku, Kyushu) stretch from north-east to south-west in a temperate zone for a distance of more than 2,000 km, with a maximum width of less than 300 km. There are many mountain ranges from 2,000 to 3,000 m high on these islands. The total land area of Japan is 377,500 km^2. The population in 1986 was 122 million, giving a density of 326 people per km^2.

Japan is located in a temperate monsoon zone. The average annual precipitation is about 1700 mm. However, the mountain ridges in the centre of the islands result in regional differences in rainfall. Thus, on Honshu the Pacific side of the island receives from 1400 to 1600 mm, while on the Sea of Japan side rainfall ranges from 2000 to 2400 mm. On Honshu, it is important to remember that it is the Pacific Ocean side of the island which has the highest population density and greatest concentration of industrial activity. Another important characteristic is that, in Japan, much of the precipitation is associated with the *baiu* and the passing of typhoons. The *baiu* (literally 'plum rain' in Japanese) occurs during a rainy season from the middle of June through to July in areas of south-western Japan. The *baiu* plays an important role in rice farming, although occasionally it comes as a heavy rain which causes landslides and floods. Typhoons migrate from the southern part of the Pacific Ocean and hit Japan between July and September. The area from Okinawa to the Kii peninsula is particularly susceptible to damages from storms and floods caused by typhoons.

Japanese rivers typically are relatively small with steep gradients. Only ten rivers are longer than 200 km. The Shinano, which flows into the Sea of Japan, is the longest river with a length of 367 km and a drainage area of

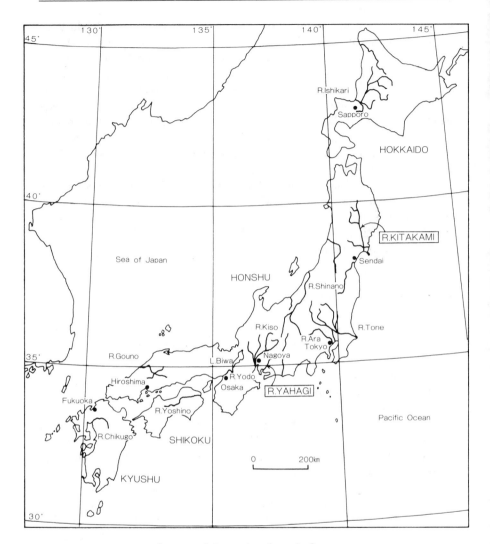

Figure 6.1   The location map of the major rivers in Japan

11,900 km^2. Both the Tone River in Kanto District and Ishikari River in Hokkaido rank with the Shinano in size. The drainage area of the former is 16,840 km^2 and of the latter is 14,330 km^2 (see Figure 6.1).

The coefficient of river regime (ratio of the maximum flow to the minimum flow) highlights the extremes of flow experienced in the rivers of Japan, thus reflecting their usually short length and their steep gradients. Rivers in Japan generally have a coefficient of river regime of more than 200 and the Tone River has a coefficient of 850. In contrast, many European rivers have a coefficient of less than 100. The Rhine has a value of

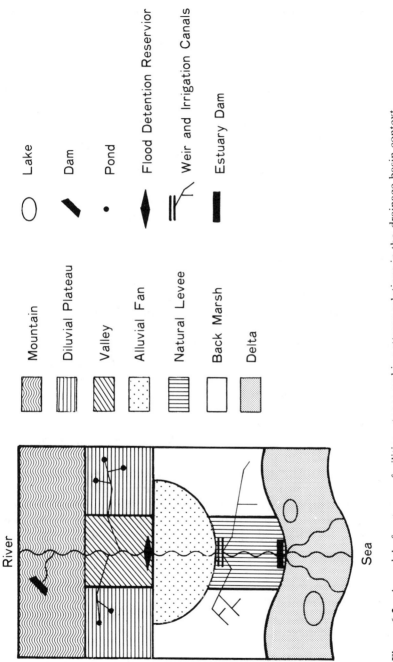

Figure 6.2 A model of water use facilities—topographic patterns relations in the drainage basin context

only 16. A high coefficient also indicates that the flow of water can fluctuate considerably over the year and that both flooding and drought can be serious problems. However, it has been difficult to construct large dams and reservoirs capable of satisfactorily controlling flows due to problems of topographical restrictions and compensation to landowners for inundation of property.

Rural land use in Japan is characterized mostly by paddy-fields which are located on the alluvial plains. *Yosui* (irrigation water) can readily be provided on the alluvial plains. In contrast, on the plateaux it is difficult to obtain river water for irrigation, creating the need for construction of ponds. During 1985, irrigated paddy-fields represented approximately 55 per cent of the 5,379,000 arable hectares in Japan.

Since the alluvial plains are created mainly by flood events, regular flooding is common. Furthermore, most of the large Japanese cities were established on the alluvial lowlands. While more recent housing and new town development have tended to be located on higher plateaux, the majority of the large cities are situated on areas subjected to flooding. This is particularly the situation for the coastal industrial zones of Tokyo, Osaka and Nagoya. When Typhoon Ise Bay occurred in 1959, the industrial zone in the southern portion of Nagoya City sustained substantial damages from a flood tide.

Groundwater is limited with regard to the alluvial plains and where it is present has tended to be overdrawn. And, as most of the river water in Japan has traditionally been allocated for paddy-fields, cities increasingly have had to look to relatively less densely settled regions for their sources of water. This has led to increasing pressure for overland transfers of water. Given the considerable fluctuations in flows in most Japanese rivers, there is usually a perceived need for structural measures to provide greater reliability of the rivers as sources of water.

Figure 6.2 shows a typical pattern of development in a Japanese drainage basin. First, in addition to an upstream dam, a downstream estuary dam is constructed. Even with such control structures, the river discharge will often exceed the design capacity of the storage measures. Thus, a flood detention reservoir is established in the central part of the basin temporarily to hold excess water. In the alluvial plain, weirs and irrigation channels have been established to take water from the river to the paddy-fields.

History of water resource development

Multiple purpose water development

After the Second World War, Japan recognized that it had an excessive population living in a country which was poor in natural resources. It was decided at that time that Japan must give priority to being able to grow its own food to provide a stable base for economic survival. In this context, water policy was guided by the Comprehensive National Land Development Act passed in 1950. The Special Area Multiple Purpose Development Projects were to be the principal mechanism under this legislation. Such projects, modelled after the Tennessee Valley Authority in the United States, were intended to ensure that development of resources was conducted in a way that was compatible with conservation of land resources.

Special areas were designated by the Prime Minister. Eighteen such special areas were created in 1951, and included Kitakami, Okutadami, Tone and Kiso. Three additional areas were designated in 1957, making 21 special areas in total. Among these special areas, emphasis was placed on hydroelectric power development at Okutadami and on comprehensive river development in Kitakami. Kitakami will be examined in more detail later.

The Special Area Multiple Purpose Development Projects had four characteristics. First, the projects were designed to increase food production and energy generation through balanced resource development and land conservation. Hydroelectric power was given the highest priority. This thrust was reinforced by the passing in 1952 of the Electric Power Development Promotion Law. Under this legislation, the Electric Power Development Company was created as a special corporation. Subsequently, nine private electricity companies were created.

Second, multiple purpose dams were a key component of Special Area Multiple Purpose Development Projects. A multiple purpose dam had been built in 1937 as part of a river water control project in Japan. It was not until the special areas were designated, however, that multiple purpose dams became a central component in water management and development. Third, another key strategy was the construction of upstream dams to facilitate production of energy and control of river flows to ease flood damages.

Fourth, the Special Area Multiple Purpose Development Projects were to be implemented for about a 10-year period. During this period, the special areas were to receive priority in central government budgeting as long as they were judged to be key districts from the perspective of national land development. Under the Comprehensive National Land Development Act, the Prime Minister could designate as special areas those 'where the resources have not been fully developed, where prevention of disasters is specially needed, and urban areas and their adjoining areas where special construction or arrangements are needed' (United Nations, 1955). In these situations, the costs of structural measures could be helped through subsidies or other measures deemed appropriate by the Government. Thus, structural measures, especially dams, were the nucleus of Japanese multiple purpose water development. Two other ideas followed.

The government concluded that a long-term and comprehensive water resource plan could not be developed without consultation with and co-operation by local communities such as the *ken* (Prefecture). To realize co-ordination with local communities, the Special Multiple Purpose Dam Law was passed in 1957. This required the Minister of Construction to confer with senior officials of other central government agencies as well as with the prefectural governors when developing water plans which included multiple purpose dams.

It was appreciated that dam construction created different benefits and costs to upstream and downstream inhabitants in a catchment basin. Upstream residents had to cope with loss of land which was inundated by reservoir waters as well as the relocation of people whose land had been lost. It was believed that the upstream inhabitants should not have to bear all of these costs on their own. Thus, through an Act on Special Measures for Reservoir Area Development in 1973 it was stipulated that the downstream inhabitants would contribute to the compensation of those adversely affected in the upper part of the catchment.

Arrangement of Three Public Water Management Laws

The impetus for the National Comprehensive Development Plan in 1962 was the rapid increase in the urban population and the increasing indus-trialization associated with economic reconstruction following the Second World War. One year before the establishment of the National Com-prehensive Development Plan, similar initiatives were taken by those responsible for water management and development. In 1961 the Water Resources Development Law and the Water Resources Development Corporation Law were passed. The former was to facilitate large-scale water use in the water system which required immediate expansion of the supply system. The latter created the agency to construct and manage water use facilities in designated water systems such as the Tone River, Yodo River, Kiso three rivers (Kiso, Nagara, Ibi), Yoshino River, Chikugo River and Ara River. The Water Resources Development Cor-poration Law absorbed the Aichi Yosui Public Corporation which earlier had been created to manage the development of the Aichi Yosui Project (*yosui*, irrigation canal).

The Aichi Yosui Project has involved four aspects: agriculture, urban, industry and power generation. The law which created the Aichi Yosui Public Corporation was passed in 1955, and allowed the control and management of water from the source to the lower part of the water system. Key components were the multiple purpose Makio Dam on the upstream portion of the Kiso River, as well as the 30,000 ha of irrigated agriculture which was the greatest national irrigated water project in

Japan. However, with the expansion of Nogoya City, in the centre of Aichi Prefecture, demand rapidly increased for both residential water and for industrial water required from construction of iron industry works in the coastal area. Consequently, with some of the agricultural area being converted to urban use, water was diverted from irrigation agriculture to urban use (Shirai, 1980). In this way, the Aichi Yosui Public Corporation was able to reallocate water use in recognition of changing conditions.

The establishment of the previously mentioned Water Resources Development Law and the Water Resources Development Corporation Law in 1961 led to enactment of the New River Law in 1964 to replace the Old River Law which had been applied since 1896. There are two important features of the new law. First, in the New River Law, water use is considered to be more important than flood control. The New River Law also relates demand for water and water rights to economic growth. In situations of drastic water shortage, drought regulations are immediately applied to those holding water rights. Second, in the Old River Law, the river was designated by the prefectural governor, but in the New River Law, rivers are classified into two categories based upon their importance to the public interest. Rivers in the first class are managed by the Ministry of Construction, while those in the second class are managed by the appropriate prefectural governor for each of the respective drainage basins. This concept has been introduced in conjunction with the integrated management of dams and reservoirs to increase the efficiency of management of drainage basins. By 1988, integrated management of dams was carried out in ten first-class river drainage basins, including the Tone River and the Kitakami River.

In reviewing the history of water resource development in Japan, it is worthwhile to mention the National Comprehensive Development Plan and its relationship to integrated water management. The first National Comprehensive Development Plan in Japan was introduced in 1962 and emphasized industrial and urban development. The Second Development Plan began in 1969 with concentration upon infrastructure development. The Third Development Plan stressed environmental management and started in 1977. None of these three plans attempted to realize integrated management of water resources. In contrast, the Fourth National Comprehensive Development Plan, which appeared in 1987, adopted integrated management of the entire water system, and also incorporated the concept of land in a 'multi-polar diffusion pattern'. This pattern is designed to avoid development concentration such as has occurred in Tokyo. It seeks to achieve a mix of development concentration ranging from low, medium and high urban area density units in a stratified yet diffused network.

Rationalization of water use

Domestic water

The total amount of water resources in Japan is about 674,900 million m^3, while the usable portion is about 88,100 million m^3, or only about 14 per cent of the total. In 1982, domestic water use represented 16 per cent, industrial water use was 18 per cent and agricultural water use 66 per cent of total use. The shift in domestic and agricultural water use has been substantial. In 1965, domestic use was only 8 per cent and its doubling by 1982 occurred at the expense of agricultural water use.

With reference to domestic water use, the Waterworks Act of 1890 was replaced by the Waterworks Law in 1977. The latter act has given responsibility for domestic water supply jointly to the Ministry of Health and Welfare and the local communities. Domestic water supply works are not conceived or developed with regard to the catchment basin but rather by individual municipalities. One result has been a deterioration of water quality in most river systems primarily because of the inadequacy of sewage treatment facilities. In 1985, 93 per cent of the population in Japan was served by a water supply system while only 59 per cent was serviced by public sewage treatment works.

Industrial water

As Japan's economic development surged forward, groundwater sources were steadily exploited. The water tables in coastal areas such as Osaka and Tokyo dropped significantly. In some places, this contributed to land subsidence which in turn contributed to flooding from high flows in rivers or high coastal tides. One response was the passage of the Industrial Water Law in 1950 which stipulated that groundwater could be drawn only from wells of specified size and that withdrawals of groundwater had to be licensed.

As groundwater sources were overdrawn or depleted, pressure grew for the overland transfers of water. In 1958 the Government passed the Industrial Water Works Law which required that local municipalities and other users could only pursue industrial water projects under the direction and surveillance of the Ministry of International Trade and Industry.

The overdrawing of groundwater, and the associated problems of land subsidence, generated calls for comprehensive management of ground and surface water. Many studies have been conducted since 1974, and diverse plans have been proposed by different ministries and agencies such as the Environment Agency, the Ministry of Construction, the National Land Agency and the Ministry of International Trade and Industry. All of these studies and proposals, however, have not led to development of a comprehensive approach. A fundamental reason for lack of progress has been the technical questions regarding groundwater which still have to be answered.

Agricultural water

A Land Improvement Law was passed in 1949 which created the opportunity for establishment of land improvement districts. The concept of land improvement districts followed the idea of water districts developed in the United States. Land improvement works associated with such districts consist primarily of agricultural water use works (dams, canals) and construction of roads and other facilities to assist agriculture. These works are carried out by the land improvement districts in conjunction with the Prefecture and the central government. Particularly large-scale projects are undertaken as national works by the Ministry of Agriculture, Forestry and Fisheries. Under a new Land Improvement Law, consolidation of cultivated lands is encouraged in order to mechanize traditional Japanese agriculture. Consolidation must be accompanied by rearrangement of roads, irrigation and drainage canals and fragmented plots if the advantages of mechanization are to be realized (Shirai, 1972).

Agricultural water is supplied under a system in which agricultural use traditionally receives priority over other uses. However, in regions experiencing rapid population growth in urbanized areas, and an associated decrease in agricultural land, mechanisms are in place to reallocate the water from agricultural to domestic or industrial users.

Water Pollution Prevention Law and arrangement of sewage works

The laws for water pollution prevention were put in place as a result of damage to the fishery in 1958 following the escape of industrial effluent from Honsyu Paper factory into the Edo River in Tokyo. The result was legislation which designated certain water areas to be kept free from pollution and which set water quality standards for industrial wastes. The various acts were consolidated in 1970 into the Water Pollution Prevention Law.

The Water Pollution Prevention Law is designed to regulate the discharge from industrial plants or other activities into public water areas in accordance with discharge standards determined through orders from the Office of the Prime Minister. The purpose of this law is twofold. First, the law is used to protect the health of the people and to maintain the environment through industrial waste water management. Second, the law is used to define clearly the responsibilities of the industrial polluter in the event of damage to human health or the environment and to set forth means for repair and compensation.

In 1967, when pollution problems were emerging as a societal concern, the Basic Law for Environmental Pollution Control was proclaimed. Under this law, water pollution was considered as one type of public pollution which also incorporated pollution of air and soil, noise, vibration, ground subsidence and odour. The government adopted the idea of preventive measures and created environmental standards for diverse kinds of pollution ranging from water to air to soil.

Most of the large cities and concentrations of industry in Japan are situated in coastal areas, resulting in a substantial problem of sea pollution in the coastal areas. In that respect, the Law Concerning Special Measures for Conservation of the Environment of the Seto Inland Sea was passed in 1973 with the purpose to conserve the environment of coastal areas. This legislation led to regulations and standards for wastes and created a system of environmental impact assessment for development proposals. The law particularly focused upon the possible problems arising from eutrophication. These laws regarding environmental pollution in general, and water pollution in particular, are administered by the Environment Agency.

When the law related to water pollution prevention was first introduced in 1958, a new sewage law was established as well. According to this law, Prefectures were authorized to deal with sewage on the basis of catchment areas. This provision was a landmark, and made a substantial improvement in conserving water quality throughout river basins. The provision of basin-wide sewage treatment facilities requires large capital investment and much discussion regarding where to site treatment works. As a result, there is much work still to be done.

Actual integrated water management

Integrated dam management of the Kitakami River basin

Kitakami River Multiple Purpose Development was the first multiple purpose river management plan in Japan which adopted the approach from the Tennessee Valley Authority (TVA) in the United States. Indeed, it is referred to as KVA. As mentioned previously, the Kitakami River was designated as a Special Area in the early 1950s under the Comprehensive National Land Development Act.

The Kitakami River is the greatest river in north-eastern Japan (see Figure 6.3). Its source is near Nanashigure-yama and Hachimantaira in the northern part of Iwate Prefecture and it flows through Iwate Prefecture and Miyagi Prefecture. The river has a main stem length of about 249 km and a basin area of approximately 10,150 km^2. The portion within Iwate Prefecture is 7,800 km^2. In 1985, about 961,000 people lived in the catchment basin.

Gorges exist in certain parts of the Kitakami River basin. The largest is

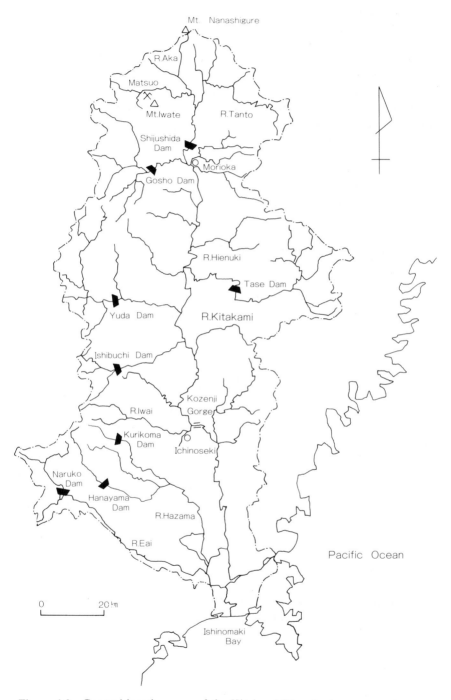

Figure 6.3 General location map of the Kitakami River basin

the Kozenji Gorge which extends 26 km from Kozenji in Ichinoseki City in Iwate Prefecture to the border of Miyagi Prefecture. Because the gorge impedes the runoff during periods of high flow, the area around Ichinoseki City has been subjected to serious flooding. In 1948, Typhoon Ione created flooding in the area around Ichinoseki City adjacent to the Iwai River, which is a tributary of the Kitakami River. This flooding caused 22,800 million yen in damages and resulted in 473 people killed or missing. In the previous year, Typhoon Katherine caused the heaviest concentrated rainfall in the history of the Kitakami River and led to the death of about 100 people.

The Kitakami Special Area Development Project (1953 to 1962) was designated by the Cabinet in 1953. When the Cabinet designation was made, 43,100 million yen were allocated for the necessary general civil works. The main components of the multiple purpose development were flood control, irrigation and power generation. Each is examined below.

The purpose of the flood control plan was to modify the design flood discharge of the main stem of the Kitakami River at Kozenji to the $9,000 \, \text{m}^3/\text{s}$ experienced during the great flood in September 1947. The means to accomplish this control were five multiple purpose dams: Tase Dam, Ishibuchi Dam, Yuda Dam, Shijushida Dam and Gosho Dam, all upstream on the Kitakami River. The five dams and the Maikawa flood detention site further downstream combine to reduce the flow reaching the Kozenji Gorge.

Downstream, in Miyagi Prefecture, the design flood discharge in the old Kitakami River has been set at $2,000 \, \text{m}^3/\text{s}$ at the estuary emptying into Ishinomaki Bay. This flow has been achieved by building dams on several tributaries of the Kitakami. Thus, on the Hazama River the discharge is restricted by construction of the Hanayama and Kurikoma dams as well as the introduction of flood detention sites in Minamiyachi and Naganuma. The Kabukuri-numa flood detention sites on the old Hazama River further control discharge. On the Eai River, the downstream flood discharge has been regulated by construction of Naruko Dam and the downstream discharge to Narusa River through a new Eai River.

The completion of the group of multiple purpose dams described above has brought real benefits through reduction in flood damages. For example, in August 1981 Typhoon no. 15 deposited the heaviest rainfall in the basin after Typhoons Ione and Katherine. However, the river level at Kozenji in Ichinoseki City was kept 2.5 m lower than during Typhoon Ione. Typhoon no. 15 resulted in the death of only one person and minor damages to houses and farm fields in Ichinoseki City.

The target for power generation from the multiple purpose dams was set at an initial target of maximum power output of 150,000 KW. In 1953 when the project was initiated, power generation in the Kitakami area of Iwate Prefecture was only a maximum power output of 40,000 KW. By 1965, power generation associated with the multiple purpose dams was at a maximum power output of 196,000 KW which is well above the originally targeted capacity.

The Kitakami area has no energy resources such as coal or petroleum, and before the project only had a little electrical energy. The hope was that the introduction of hydroelectric power would stimulate mining and manufacturing activity in the region. Unfortunately, the introduction of power has had little impact on mining or manufacturing. However, this power has contributed substantially to the electrification in urban areas and in farm villages.

The irrigation activity associated with the development project has been significant. The Kitakami area has a relatively low annual precipitation, ranging from 1100 to 1200 mm, and often experiences drought during the critical growing periods of rice. With unreliable water supply and constraints caused by topography, much land remained in forests and pasture rather than in cropland.

After the multiple purpose dams and associated irrigation canals were constructed, 11,500 ha of new paddy-field and 8,100 ha of plowed field in Iwate Prefecture were added to the cropland in the area. Moreover, a supplemental water supply for the existing paddy-fields covered about 22,200 ha in Iwate Prefecture and 17,100 ha in Miyagi Prefecture. This supplemental water has helped the farmers as it has reduced the uncertainty about adequate water during the critical period in which the rice-fields require irrigation. Through the expansion of both the rice crop and breeding of milk cows, the local economy was changed significantly (Shimpo, 1976).

Another innovation occurred in relation to the expansion of paddy-fields. In Japan it had been said that paddy-fields could not be created on volcanic ash soil due to its very great water requirements. In the reclamation project at Mt. Iwate, a crushing and compaction method to treat volcanic ash soil was developed by the Department of Agricultural Engineering at the University of Iwate. Without this technological innovation, the benefits from the provision of a reliable supply of irrigation water would not have been nearly so great for the local farmers. On balance, the irrigation benefits to the area have been large. The villages at the base of Iwate Mountain had been relatively poor, and the provision of water has been a major benefit for them.

It may be questioned whether the maximum benefits from flood control, power generation and irrigation have been realized. The main reason for this question is that there is no strong authority which integrates these multiple purposes as flood control is managed by the Ministry of Construction, and power generation is the responsibility of the Ministry of International Trade and Industry. Irrigation and agricultural development are in the domain of the Ministry of Agriculture, Forestry and Fisheries.

In an attempt to overcome the divided responsibility regarding the different management purposes in the catchment, the Ministry of Con-

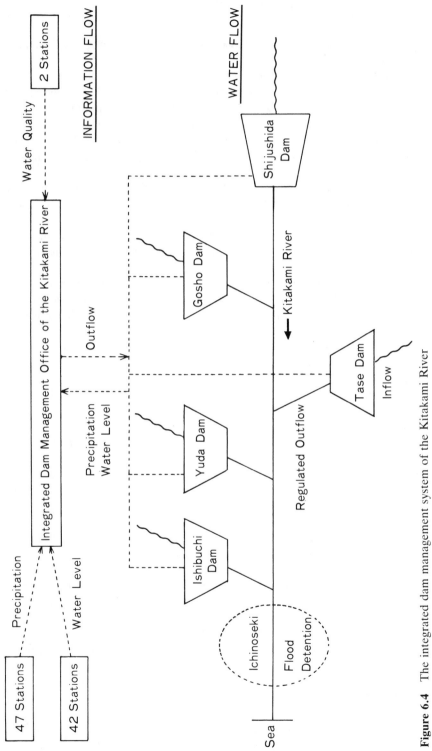

Figure 6.4 The integrated dam management system of the Kitakami River

struction established the Integrated Dam Management Office of the Kitakami River Basin (IDMK) whose task is to manage in an integrated manner the flood control, water use and water quality aspects of each dam both individually and as part of an interconnected group. In this work, the Ministry of Construction works co-operatively with the Iwate Office of the Thohoku Regional Construction Bureau which has been in charge of river works. The integrated management system for the Kitakami multiple purpose dams is shown in Figure 6.4.

The five major dams have slightly different purposes. On the main stem of the Kitakami River, the Shijushida Dam was designed for flood control and generation of hydroelectric energy. The other four main dams are on tributaries of the Kitakami River and include the following purposes: Gosho Dam (flood control, hydroelectric power, irrigation, domestic water use) was the first assigned dam under the Act on Special Measures for Reservoir Areas Development; Yuda Dam (flood control, hydroelectric power, irrigation); Ishibuchi Dam (flood control, hydroelectric power, irrigation), the first rock fill dam in Japan; Tase Dam (flood control, hydroelectric power, irrigation).

For flood control the IDMK draws upon the 47 precipitation stations and 42 water gauging stations to develop flow forecasts upon which dam operations are based. The multiple purposes expected of the dams can create problems and conflicts for the managers, as illustrated by flood control, power generation and irrigation.

For flood control purposes, the reservoirs should be drawn down to minimal levels prior to the arrival of the flood season associated with typhoon attacks between the beginning of July to the end of September. To draw down the reservoirs completely, however, would jeopardize production of hydroelectricity and the release of water to accommodate the irrigation needs of the paddy-fields. In the case of drastic water shortages, IDMK intervenes to attempt to avoid or minimize conflict between generation of hydroelectricity and irrigation.

Water quality has also been an issue, since the Kitakami River is affected by the escape of strong acidic wastes from a sulphuric mine on the Aka River, a tributary in the upper part of the catchment. As a result of the release of acidic wastes in the upper part of the basin, the water stored behind the Shijushida Dam on the main stem is not used for irrigation or domestic use.

Even after the Matsuo sulphuric mine stopped operating, pollution continued in the upper portion of the Kitakami River. Consequently, the Ministry of Construction has been introducing calcium carbonate into the river to neutralize the acid wastes. Notwithstanding these efforts, there remains polluted sludge in the bottom of the Shijushida reservoir which continues to be a problem. The IDMK works with the Kitakami River Pollution Council to try and improve water quality in the river so that it will meet the standards established by the Environment Ministry in 1973.

The design flood discharge of the Kitakami River at Kozenji has been modified from the 9,000 m^3/s to 13,000 m^3/s to accommodate rapid urbanization in this basin. To meet this new discharge standard, further structural and non-structural measures are being or will be taken in the basin. Three additional dams are scheduled to be built on three branches (Tanto River, Hienuki River, Iwai River). A new Ishibuchi Dam will be built. The Maikawa flood detention sites are being expanded and Ichinoseki flood detention sites are under construction. These flood detention sites will be used for agriculture as well. The Ichinoseki flood detention sites are the largest in Japan with a flood detention area of 1,450 ha and a capacity of 76,800,000 m^3. Their construction will require the removal of 416 houses. In 1984, 218 houses were moved to higher areas not subjected to flooding. The movement and resettlement of people has taken considerable time to resolve because of the need for compensation (Uchida, 1985).

Yahagi River and river basin management

As demonstrated during the discussion of the Kitakami River, the Japanese administrative system uses different agencies for flood control, power generation and irrigation. There is no single organization with basin-wide authority for a number of management functions which are interdependent. Since inhabitants of the basin cannot expect to find a single agency which takes a comprehensive view of the entire catchment, they have turned to quasi-government organizations to carry out this task. The Yahagi River Basin Water Quality Protection Association (YWPA) was the first quasi-government organization created in Japan to consider basin-wide problems in a comprehensive manner (Naitoh, Sugimoto and Harashima, 1988).

The Yahagi River is situated in the central part of Aichi Prefecture (see Figure 6.5). The river is about 122 km long and the basin has an area of 2,264 km^2. In 1985, about 1,259,000 people lived in the catchment. The Meijiyosui irrigation canal from the Yahagi River presently serves 7,660 ha of rice-fields with the water channel network provided with water from the Habu and Yahagi dams. Furthermore, given the high degree of urbanization, the main channel of the Meijiyosui has been converted into a pipeline and covered, with the upper portion now used as a roadway. The management and maintenance of the irrigation system and fields are done by the Meijiyosui Land Improvement District, consisting of some 15,000 members. In addition to rice farming, the Meijiyosui Land Improvement District developed considerable horticulture due to its location in the Chukyo Industrial Zone—to the extent that prior to the Second World War it was referred to as the 'Denmark of Japan'.

There are two major sources of water pollution in the Yahagi River. One source has been the growing urbanization and industrialization represented

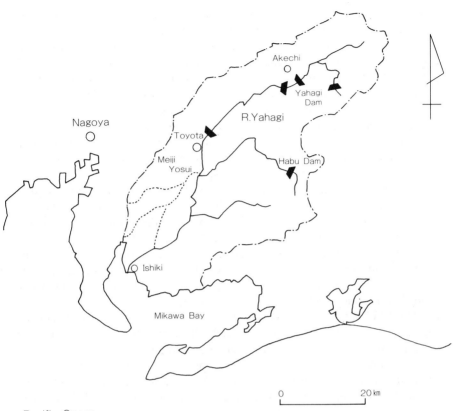

Figure 6.5 General location map of the Yahagi River basin

by the homes developed by the Japan Housing Corporation and the large-scale automotive plants built by Toyota Motor Corporation. A second source is the weathered granite soil found in the mountainous upstream area of the Yahagi River system. This soil is well suited for ceramic and construction material, and its extraction has led to the discharge of a white sludge into the river.

The pollution in the Yahagi River was steadily worsening, and in 1967 the Meijiyosui Land Improvement District representing the farmers began to examine possible remedial actions. Further interest in such action arose from damage from the pollution on the sweetfish in the river and on cultivation of seaweed in the downstream estuary. The growing concern from a number of interests led to the establishment in 1969 of the Yahagi River Basin Water Quality Protection Association (YMPA).

Eighteen groups collaborated to form the YWPA. These included agricultural groups, fishery co-operative associations and municipalities in the lower portion of the Yahagi River which were suffering from the deteriorating water quality. A YWPA office was established with the Meijiyosui Land Improvement District, and Mr Renji Naitoh was appointed as Secretary General. The YWPA has expanded and by 1987 involved 49 groups (four agricultural, 19 fishery, 25 municipalities and one Prefecture) covering the entire basin and supported by numerous other groups. It has developed into an organization concerned with river basin management in general, and deals with the environmental administrations of the municipalities. The YWPA has two basic objectives: (1) maintenance of water quality according to established standards; and (2) assessment of the appropriateness of large-scale developments.

An index of quality has been adopted as a simple and readily understandable measure of water quality in the river. It was decided that the standards developed under the Water Pollution Prevention Law or by the Prefecture were not sufficient, so the YWPA created more demanding standards. For example, with respect to biological oxygen demand, the national and Prefectural standards are 160 (120) and 25 (20), respectively [units in ppm, with the number in parentheses denoting the allowable limit], while the YWPA standard is 10 ppm.

The YWPA's second function relates to large-scale development involving more than 20 hectares. Within Aichi Prefecture there is no law related specifically to environmental assessment, although several other laws provide general guidelines for land development. The Aichi Prefecture has thus stipulated that consent from the YWPA should be obtained for large-scale developments before the Prefecture will give approval. This arrangement has been in place since 1980.

The YWPA has developed a four-stage procedure regarding provision of its consent. This includes preparation of an environmental assessment, review and approval of YWPA, initiation of environmental monitoring and use of a Council on Environmental Pollution Prevention (see Figure 6.6).

For large-scale development, YWPA requires that an environmental assessment be prepared. YWPA believes that preparation of assessments has two effects. First, the proponent and contractor are made more aware of the consequences of their activities on others in the catchment. Second, the information provided in the assessment allows the YWPA to reach an informed decision about the appropriateness of the proposed development. YWPA is particularly concerned that the assessment addresses the implications of the development for water use, water quality, flood control and transportation.

Since the YWPA represents such a cross-section of interests in the catchment, it has been accepted that YWPA is an appropriate organization to review the environmental assessment and give approval or request further study. Only after YWPA has accepted the environmental assessment will Aichi Prefecture give permission for the development to proceed.

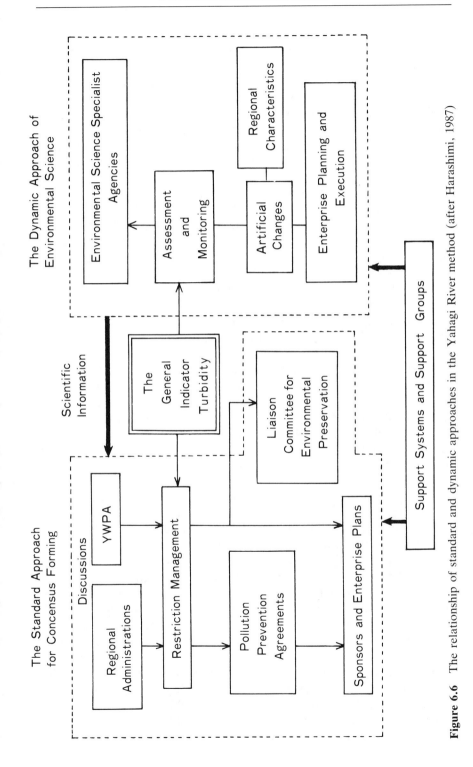

Figure 6.6 The relationship of standard and dynamic approaches in the Yahagi River method (after Harashimi, 1987)

After receiving approval, but before beginning construction, the proponent must arrange for environmental monitoring to begin. The monitoring procedure was developed by Taiyo Kiko Co. Ltd. and provides a range
of diverse measures relating to the water system and the entire ecosystem.
This computerized monitoring system provides data on a monthly basis.

In order to ensure communication among the proponent, the contractor,
the municipal governments and YWPA regarding the implications of the
environmental monitoring, a forum was created to bring all of these
interests together. The Council on Environmental Pollution Prevention is
organized by senior officials from the municipalities, and includes super-
visors from the Prefecture, representatives from the municipalities, the
proponent and contractor, YWPA representatives as well as specialists on
environmental matters. The Council usually meets once a month, but may
sometimes only meet once every two or three months. The Council reviews
the results of the environmental monitoring, and if problems emerge, can
direct the proponent or contractor to take necessary mitigative action.

It has been found that the cost of using this four-stage procedure is about
1 per cent of the total cost of large-scale projects and about 2 to 5 per cent
of the construction costs. In addition to the four-stage procedure, three
other activities have been created to support the approval and monitoring
process. First, in 1970 the Yahagi River Basin Development Research
Association was established. This research organization was created as a
co-operative venture by 26 municipalities in the catchment and the first
director was Dr Gohei Itoh, a geographer and President of Aichi University of Education. The purpose of the research association was to ensure
that individuals working in the municipal environmental departments were
aware of latest developments in environmental research in general and of
environmental work done in the catchment.

A second initiative was to establish connections between upstream and
downstream municipalities in the Yahagi River basin. For example, in
1977, a sister town affiliation was formed between Akechi-cho, a town
specializing in ceramics in the upper part of the basin, and Ishiki-cho, a
fishing town in the lower part of the basin. Through exchanges of school-
age children and information about the economic activities of the two
towns, residents were able to appreciate better the relative importance of
the river to the two communities. The third element involves the development of general environmental education within the catchment. Schoolchildren are shown how water quality testing is conducted, and in the
process their awareness of the importance of water quality in the basin is
enhanced.

The concept of comprehensive river basin management which has
evolved in the Yahagi River basin follows the policy in the Third Comprehensive National Development Plan initiated in 1977. The Third Plan seeks
to achieve harmony by maximizing economic growth while simultaneously
minimizing environmental degradation. This involves the creation of an

integrated 'residence' policy for river basins. Such a policy views any river basin as a community sharing a common destiny. In order to benefit, everyone, upstream and downstream, must *communicate* with one another. Furthermore, the use of the monitoring system of Taiyo Kiko Co. Ltd. and the creation in 1986 of the Yahagi River Environmental Technology Research Association, the research section of the Taiyo Kiko Co. Ltd., has ensured that environmental science underlies many of the development and management decisions. The experience in the Yahagi River basin also demonstrates the impact which a non-government organization can have in river basin management. It shows that effective comprehensive river basin management does not have to depend solely on central government agencies.

Conclusions

In Japan, responsibility for water management and development is shared by different agencies such as the Ministry of Construction, Ministry of Agriculture, Forestry and Fisheries, Ministry of Health and Welfare, Ministry of International Trade and Industry, National Land Agency and the Environment Agency. A similar division of responsibility is found at the local level of government.

It is only recently that the constraints of inadequate water quantity or quality have been appreciated in Japan. Although Japan has many laws relating to water, there is no basic law which stipulates that a comprehensive approach should be pursued in the context of river basins. The primary concern to date has been upon flood control, water quality, droughts and hydroelectric power production. Problems have tended to be local rather than national in scope which has reinforced the tendency not to consider water management through either a comprehensive or an integrated approach. Some exceptions to this generalization exist, however, as illustrated by the experiences in the Kitakami River and the Yahagi River. Furthermore, the National Comprehensive Water Resources Plan formulated in 1987 should counter the tendency to approach water management in a less than comprehensive or integrated manner. If a watershed-based comprehensive approach were to be developed, three considerations will require attention.

First, any law to guide comprehensive water development and management must recognize the integrity of the hydrological cycle. In other words, the law should recognize the interrelationships among ground and surface water, water quantity and water quality, and water and related land-based resources. If the interconnections of the water cycle are arbitrarily divided up by the legal system then it will be difficult to realize a comprehensive approach.

Second, in water management it is inappropriate to focus exclusively upon the water within the river banks. It is necessary to remember that if multiple purpose projects are to be used it will be necessary to think not only of the water in the rivers but also about the water flowing overland to enter the rivers as well as about the activity on the adjacent land.

Third, it should not be forgotten that water management is a means to an end, rather than an end in itself (Mitchell, 1986). The ultimate end surely is to improve the quality of life of the people in the river basin while maintaining the integrity of the natural environment. If water managers keep this idea in mind, then it should be possible to achieve the 'sustainable development' which has been advocated in *Our Common Future*, the report from the World Commission on Environment and Development (1987). The systematic use of the ideas provided in the National Comprehensive Water Resources Plan of 1987, combined with the suggestions offered here, should provide a way to improve both water management and quality of life in Japan.

References

Craine, L. E., 1969, *Water Management Innovations in England*, Johns Hopkins Press, Baltimore.

Easter, K. W., Dixon, J. A. and Hufschmidt, M. M. (eds.), 1986, *Watershed Resources Management—an Integrated Framework with Studies from Asia and the Pacific*, Westview Press, Boulder.

Mitchell, B., 1986, 'Integrated river basin management: Canadian experiences', *Hydrology and Water Resources Symposium 1986: River Basin Management*, Institution of Engineers, Australia, Barton, A. C. T., pp. 140–7.

Naitoh, R., Sugimoto, M. and Harashima, R., 1988, *Consideration of Management Systems and Concensus Forming in River Basin Management—the Development of the 'Yahagi River Method'*, Report presented at the First Expert Group Workshop on River/Lake Basin Approach to Environmentally Sound Management of Water Resources, 8–19 February 1988, Otsu and Nagoya, Japan.

National Land Agency, 1987, *National Comprehensive Water Resources Plan— Water Plan 2000*, Tokyo.

Okamoto, M., 1983, 'Japanese water transfer: a review' in A. K. Biswas, Z. Dakang and J. E. and L. Nikum (eds.), *Long-distance Water Transfer: A Chinese Case Study and International Experiences*, Tycooly International Publishing Limited, Dublin, pp. 65–75.

Porter, E., 1978, *Water Management in England and Wales*, Cambridge University Press, Cambridge.

Shimpo, M., 1976, *Three Decades in Shiwa—Economic Development and Social Change in a Japanese Farming Community*, University of British Columbia Press, Vancouver.

Shirai, Y., 1972, *Nihon no kochiseibi* (Land improvement in Japan), Taimeido, Tokyo.

Shirai, Y., 1980, *The Water Use and the Development of the Region Involved in the*

Aichi Irrigation Project, the 24th International Geographical Congress, Main Session Abstracts, Vol. 3, 1–5 September 1980, Tokyo.

Shirai, Y., 1987, *Suirikaihatsu to chiikitaio* (Water-use development and regional responses), Taimeido, Tokyo.

Uchida, K., 1985, *Yusuichi to chisuikeikaku* (Flood detention sites and flood control planning), Kokonshoin, Tokyo.

United Nations, Economic Commission for Asia and the Far East, 1955, *Multiple-purpose River Basin Development*, United Nations, New York.

World Commission on Environment and Development, 1987, *Our Common Future*, Oxford University Press, Oxford.

CHAPTER 7

Water resources and management in Poland

Zdzislaw Mikulski

Introduction

The integration of water management in Poland has been attempted since 1945 but has not yet been satisfactorily realized. This chapter describes the general state of the water resources and their characteristics in Poland. Special attention is devoted to the first Polish national plan of water management, elaborated in 1952–5, and the prospects for its development. The process of organization of water management, and changes in the conception of that organization, are discussed as well as the development of integration forms and the co-ordination of water management in Poland. It is believed that there is a need for further organizational changes in the management of water resources which should be based on administrative units within natural hydrographic boundaries. The main future tasks are substantial improvement of the purity of water and future development of facilities for water storage.

Water resources in Poland

Nearly the entire Polish territory lies within the drainage basin of two major European rivers—the Vistula and the Oder—both draining north to the Baltic Sea. The water balance in the country consists of: precipitation (P), evaporation (E), river runoff (surface H′ and underground H″), and retention in deeper aquifer layers which is not included in the water balance.

The calculation of the water balance in Poland is based on the following equation: $P = (H' + H'') + E$. Following L'vovich (1979), the total moisture of the area, or the soil humidity, is identified as W and is calculated as $W = P - H' = E - H''$. Using this procedure, the water balance for Poland has been calculated by Gutry-Korycka (1985) in the

172

Department of Hydrology of the Faculty of Geography and Regional Studies at Warsaw University.

Poland is supplied with precipitation from snow in winter and rain in summer. The quantity of precipitation depends mainly on the maritime air masses, most frequently coming from the north-west, and on continental air masses from the south-west. Two climates, maritime and continental, collide over Poland. The greatest part of precipitation falls in the summer. The portion of snow in the total annual precipitation is not more than about 15 per cent. The greatest amounts of precipitation occur in the south (the Carpathians and the Sudetes) and in the seaside belt of the Baltic Sea. The mean annual precipitation in Poland amounts to 716 mm. The country's moisture (W) is an important element in water management for agriculture. Its mean annual value is 640 mm. As with precipitation, higher moisture values occur in the south and north of Poland.

The regional evaporation (E), assumed to be the difference between precipitation and river runoff, is the third most important element in the Polish water balance. Its mean annual value is 545 mm, being somewhat higher in the mountains in the south.

From the viewpoint of Poland's water resources, the most important water balance element is the river runoff (H'). Its mean annual value is about 170 mm. It is highly irregular, the result of both natural conditions (the runoff process of the main rivers in the mountainous and upland regions) and of unsatisfactory management of the rivers. The outcome is frequent periods of low runoff in winter, mainly in the mountains, and in summer and autumn, mainly in the lowlands. It also results in floods, which in the mountains are the result of heavy rains and in the lowlands due to melting of snow over vast areas. High values of river runoff prevail in the mountains and to a certain extent in the coastal zone. On average, however, Poland is poorly endowed with water.

In the calculations of Poland's water balance, the river runoff is divided into surface and underground components. Detailed calculations for all major rivers have shown that the underground runoff (H″) includes 55 per cent of the country's total runoff (95 mm). Thus, the surface runoff constitutes 45 per cent (76 mm). This distribution between surface and underground runoff partially compensates for the irregularity of the surface runoff. The main aspects of the water balance are presented in Figure 7.1.

From Figure 7.1 it is apparent that river runoff amounts to about 54 km³ annually. This is an average value, and in dry years the runoff may fall to 70 per cent of the mean value (38 km³). The necessity of preserving a certain amount of runoff for biological or ecological reasons, referred to as inviolable runoff, has resulted in the guideline that the mean low runoff in river systems should be maintained. Thus, in estimating available water resources, engineers consider the difference between gross runoff and the inviolable runoff.

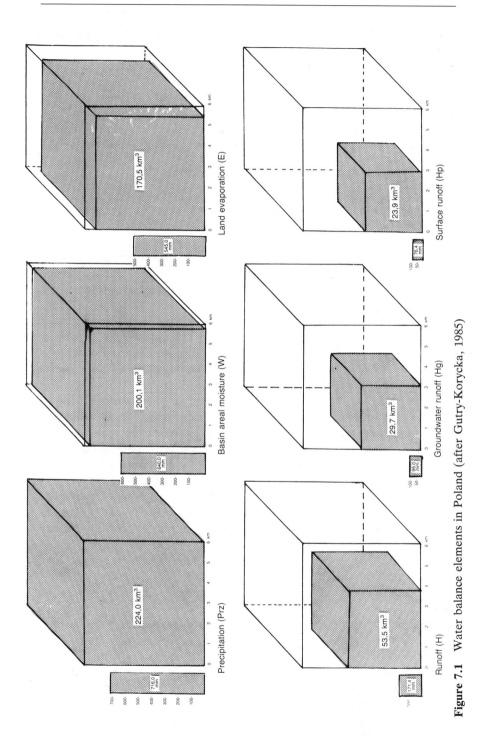

Figure 7.1 Water balance elements in Poland (after Gutry-Korycka, 1985)

Water management in Poland

Developing a national water plan

Important social and economic changes as well as Poland's new territorial shape after the Second World War resulted in a new approach to water management. The Polish People's Republic entered the path of a planned socialist economy following a three-year plan (1947–9) for restoration of the country from war damages. The new State Committee for Economic Planning, founded in 1949, was among other things to deal with problems of water management. One of the first tasks of this Committee was the development of a national economic plan.

In 1951, the Polish Academy of Sciences (PAN) was established as the organization responsible for scientific development. As part of PAN, a committee on water management was created which began working on the country's first water management plan for the period extending to 1975.

In preparing this first water management plan, assumptions were made about the future population in Poland. Later, it became clear that the future population had been overestimated, and it was assumed that Poland would have 38 million people in 1975.[1] Assumptions were also made regarding intensive development of industry and associated urban growth which were expected to result in increased demand for water. All aspects of water use were considered within the water management plan relative to the anticipated economic development of the country. However, little attention was given to the possible increase in water pollution in the country as a result of industrial and urban growth.

The 'Outline of a Perspective Plan for Water Management in Poland', prepared by the PAN in 1956, contained general statements concerning the water situation and directions for water management in Poland until 1975. In this plan, it was concluded that Poland's water resources, in the form of river flow, were sufficient to cover water demand. However, due to substantial variability in river flow in time and in space as a result of the occurrence of thaw and rain-fed floods as well as periods of drought, it was suggested that greater attention should be given to the construction of storage reservoirs in the upper tributaries of the Vistula and Oder rivers, two of the main Polish rivers flowing from the Carpathian and Sudeten Mountains to the Baltic Sea. In addition, a general improvement of water management in mountainous regions was recommended.

As a result of the uneven distribution of water resources in Poland, it was suggested that industry should be decentralized. It was also believed that surface water should be relied upon to supply industries and cities, with groundwater being kept as a reserve for the future.

Any significant deterioration of water quality was not expected. The construction of reservoirs which could be used to dilute wastewater concentration in rivers was seen as appropriate action to counter any excessive pollution. In contrast, an increase in water demand for irrigation was predicted as well as an expansion of forested areas. The expansion of forested areas was considered desirable to improve water management.[2] Anticipated increases in demands for energy highlighted the contribution which hydroelectricity could make, and subsequently led to water being thought of as 'white coal'. Finally, water reservoirs and barrages were viewed as essential to facilitate the expansion of inland navigation which would support industrial expansion. The lower course of the Vistula River was to be canalized both for hydroelectricity and for navigation purposes.[3]

The *Perspective Plan for Water Management in Poland to 1975*, the first one in the country and one of the earliest ones in the world, became the guideline for water management in Poland for several years. In spite of its general character, it identified the main problems, suggested solutions and indicated fundamental directions for action related to the economic and social development of socialist Poland (Mikulski, 1982).

Organization to implement the national plan

When the long-term plan of water management was prepared in the first half of the 1950s, water management problems in Poland were shared among three government ministries: Municipal Economy, Agriculture and Navigation. Within the Ministry of Navigation, the Central Management of Inland Waterways department was established in 1952 to supervise local executive organizations for district management of waterways in several larger cities such as Warsaw, Wroclaw, Poznan and Cracow as well as numerous local waterways. Management offices for large capital projects were established in Cracow, Wroclaw and Warsaw as well as a design office known as 'Hydroprojekt' and later a state enterprise called 'Hydrogeo', with the latter specializing in civil engineering. Thus, it can be seen that water management was very fragmented.

This situation was not conducive to the development of integrated water management in general nor to the development of large projects. Meanwhile, the implementation of the Six Year Plan of Economic Development and Building of the Principles of Socialism in Poland between 1950 and 1955 was concluded. This brought distinct increases in investments and considerable expansion in industrial production. Many new water investments and projects, accompanying the new industry, came into existence. For example, in a reach of the Upper Vistula River, the large Goczalkowice reservoir was built to serve as a source of water supply for Silesia, the major Polish industrial region. However, the new industrial development also resulted in a sudden increase in pollution of water, air and soil.

In order to implement projects under the *Perspective Plan for Water Management in Poland*, which was accepted and confirmed by the Government in 1957, it was decided to create a new department with responsibility for all water problems. Thus, in 1957, the Ministry of Navigation and Water Management was established. Among other things, this Ministry was to be responsible for: (1) complex long-term planning for water management; (2) water protection, taken over from the Ministry of Municipal Economy; and (3) issuing legal licences connected with the use of surface and groundwater, taking over from the Ministry of Agriculture.

During 1958, departments of water management were established in provinces and administrative districts. In 1960, the Ministry of Navigation and Water Management was replaced by the Central Department of Water Management which was given responsibility for the planning and co-ordination of water management. Water problems were at this time considered together with problems of air quality and other aspects of environmental quality. The Central Department of Water Management soon formulated a new water management plan to the year 1980, with more general considerations extending to the year 2000. This plan and subsequent ones included consideration of increases in water use due to rapid industrial expansion, and emphasized the necessity for water pollution control.

An urgent need was a new approach for the laws and regulations concerning inland water conservation. The first Polish law regarding inland waters was passed in 1922, shortly after Poland regained independence after the First World War. The new social and economic situation after the Second World War required different legislation which would meet the requirements of a socialist economy.

A group of scientists representing various disciplines and professions began work on a new law through the Committee of Legal Sciences of the Polish Academy of Sciences. Experiences in other countries, particularly those in socialist countries, were reviewed and evaluated. In 1962 the SEYM (Parliament) of the Polish People's Republic passed the Water Law bill, which from then on was the state law which regulated management of water. In 1974, after twelve years of operating under this legislation, a new law was passed (Water Law 1979). The new law had eight sections: (1) general regulations; (2) water utilization; (3) water protection against floods; (4) hydroengineering (including land improvement and water supply); (5) local water authorities; (6) water rights registers and cadastral survey of water management; (7) fines and taxes; and (8) temporary and final regulations. This law is still in force despite numerous organizational changes for water management and territorial changes in the adminis-tration of Poland.

Further organizational change

Organizational changes for water management in Poland continued during the 1970s. In 1972, the Central Department of Water Management was disbanded and its activities were transferred to the Ministry of Agriculture. At this time, the organizational structure for water management once more resembled the situation which existed in the mid 1950s.

Unfortunately, soon after this reorganization an economic recession began in Poland, reaching its climax at the beginning of the 1980s. During this recession, the implementation of water management programmes in Poland broke down. Meanwhile, substantial increases in water use, initially noticed in the 1960s and 1970s, made officials realize that water could become a barrier to the economic development of the country in the first quarter of the twenty-first century. It was estimated that by the year 2000 water use would equal the magnitude of surface river flows in dry years. This realization induced a different attitude towards water management from many Polish scientists, an attitude that accepted the need for a systematic approach to water management.

In the mid 1970s, Governmental Research Development Programmes were introduced to conduct studies in major economic and social areas. Eight programmes were formulated, one being entitled 'Formation and Exploitation of Water Resources'. At first these programmes were planned for the five-year period of 1976 to 1980, but later were extended for another five years (1981 to 1985). The water resources programme had an important role, especially in the development of scientific foundations for contemporary management of water resources in Poland. It initiated broad hydrologic research, development of mathematical modelling methods in hydrology and the introduction of systematic methods for water management.

During the second five-year period (1981–85), two pilot water management systems were introduced: (1) in the industrial region of Upper Silesia in southern Poland and (2) in the agricultural region of the upper Notec River in north-central Poland with its fully developed chemical and agricultural food industry (Mikulski, 1982). Both pilot systems were being examined to determine the most effective way to management water resources. Unfortunately, the social and economic crises of the early 1980s limited the effectiveness of these pilot exercises. Nevertheless, water management in Poland should gain much experience from both of these pilot ventures relative to future water system management on a national scale.

The doubling of water demand between 1960 and 1970 caused by excessive industrial investments with accompanying increases in the amount of wastes and further degrading of water quality, were the reasons for the introduction of charges for specific water consumption and sewage disposal. A Fund for Water Management was created in order to support the scanty capital budget assigned for more important water projects (reservoirs and barrages or weirs), regulations and hydrostructures on rivers and streams, flood protection and water pollution control, scientific study and design work.

During the first five-year period of 1976 to 1989, the Fund for Water Management collected about 20,000 million zlotys. Approximately 25 per cent of this amount came from payments for water consumption and 75 per cent from payments for sewage disposal. The accumulated funds made over 50 per cent of the budget outlay assigned for the development of water management and protection during this period. It is difficult, however, to estimate unequivocally whether the charges altered water consumption and sewage disposal. A further complication in estimating the impact of the charges was the deep economic crisis which arose shortly after their introduction. This crisis undoubtedly influenced water use and waste disposal patterns.

Patterns of water use

On the basis of data from the Central Statistical Office, a rather intensive increase in water demand in industry occurred from the mid 1950s, the time at which the six-year plan for the industrialization of Poland was concluded. In 1960, the total water demand reached 5.5 km^3 including 3.8 km^3 in industry (including thermal power energetics). The share of industry in total water consumption reached 69 per cent. Relatively small water demand occurred in the municipal economy (0.9 km^3 or 16 per cent) and in agricultural irrigation (0.7 km^3 or 12.7 per cent). A peak water demand occurred in the second half of the 1970s when 14.5 km^3 was consumed in one year, including as much as 10.5 km^3 in industry, or nearly 73 per cent, with a small increase in municipal use (2.4 km^3 or 16.6 per cent) and also in agriculture (1.5 km^3 or 10 per cent).

The economic crisis at the end of the 1970s and the beginning of the 1980s caused a decrease in water demand for several years compared to 1978. It was not until 1983 that a slight increase was observed again, to be followed by another decrease in consumption. Generally, a small increase in water use occurred over the years 1981 to 1987. Throughout this period, industrial water use has dominated with industry's share remaining at about 70 per cent. Municipal use has ranged from 18 to 19 per cent, and agricultural use has been about 11 per cent. In absolute terms, water consumption in 1987 was about 11 km^3 by industry, 3.0 km^3 by municipalities and 1.7 km^3 by agriculture (see Table 7.1).

The Perspective Plan for Water Management in Poland, prepared in the second half of the 1970s, provided for much more intensive use than previously had been anticipated to the year 2000. By 2000, water use was estimated to be 44 km^3 and at that level would exceed the total river flow in a dry year. The distribution of demand among various sectors of the economy was also predicted to change. Industrial use was expected to drop to 54 per cent of the total. It is now known that this prediction was incorrect. A small increase in the share of the municipal sector had been

Table 7.1 Annual water usage in Poland, 1960–87

Users		1960	1970	1978	1979	1980	1981	1982	1983	1984	1985	1986	1987
Industry /including energetics	km^3	3.8	6.9	10.5	10.0	10.1	10.0	10.2	11.3	11.2	10.9	10.8	10.8
	%	69.1	68.6	72.9	70.6	71.5	70.6	69.1	70.5	70.7	70.7	69.3	70.0
Muncipal sector	km^3	0.9	1.5	2.4	2.5	2.7	2.7	2.8	2.9	2.9	2.9	3.0	3.0
	%	16.4	14.8	16.6	17.9	19.2	19.3	18.9	18.0	18.0	18.9	19.2	19.6
Agriculture /irrigation	km^3	0.7	1.7	1.5	1.6	1.3	1.4	1.8	1.8	1.8	1.6	1.8	1.7
	%	12.7	16.6	10.5	11.5	9.3	10.1	12.0	11.5	11.3	10.4	11.5	10.7
Total	km^3	5.5	10.1	14.5	14.2	14.2	14.2	14.7	16.0	15.9	15.4	15.6	15.5

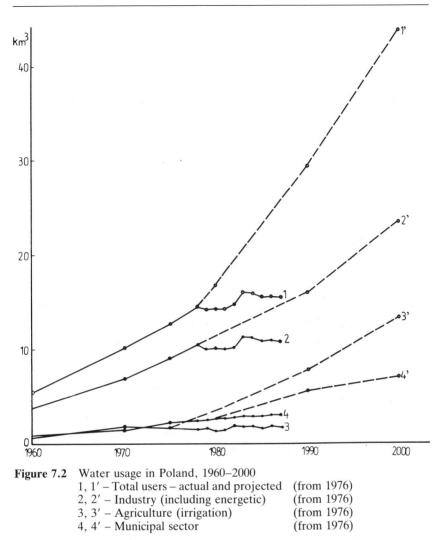

Figure 7.2 Water usage in Poland, 1960–2000
1, 1' – Total users – actual and projected (from 1976)
2, 2' – Industry (including energetic) (from 1976)
3, 3' – Agriculture (irrigation) (from 1976)
4, 4' – Municipal sector (from 1976)

forecast as well as a large increase (to 30 per cent) for agriculture (see Figure 7.2). These forecasts were also incorrect.

Addressing water pollution problems

During 1981, the scientific and technical staff in the field of water resources management in Poland made efforts to create an independent department which would be able to counteract the general degradation of water quality and to ensure the preparation of water management strategies to support economic development. In 1983, the Department of Environmental Protection and Water Management came into being. It was concerned with (1) environmental protection, including water protection, (2) water management, (3) preservation of rivers and streams, (4) waterways and (5) meteorologic and hydrologic staff. In 1985 the Department was replaced by the Ministry of Protection of Environment and Natural Resources. In this manner, water management became independent from central government economic agencies. At this stage it was hoped that an appropriate organizational structure had been created to deal with water management into the twenty-first century.

Designing for integration and co-ordination

When the reform of territorial administration was completed in 1975 and 49 provinces were formed, jurisdiction for water management was given to the provinces. In the provincial offices, Sections of Environmental Protection, Water Management and Geology were established, as well as provincial committees for flood protection. These sections could grant Water Law licences in their regions. The administration, control, conservation and exploitation of inland waters and rivers, navigable canals and lakes fell within the jurisdiction of seven Directorates for Water Management. However, the areas of the water management Directorates do not coincide entirely with river basins or catchments (see Figure 7.3).

The implementation of the existing water management plans with respect to national level investments is the responsibility of the Ministry of Protection of Environment and Natural Resources. Implementation of more localized investments is the responsibility of local authorities, provinces, or economic departments. Multipurpose water management projects are financed from central government funds whereas minor projects are financed by the local authorities. Furthermore, a continuing problem in managing water resources is co-ordinating various levels of government since state and catchment boundaries do not coincide.

Th 'General Conception of the Organization Model of Water Management in Poland' (General conception . . . , 1987), developed in 1986 by an engineering lobby group, indicated fundamental principles for management of inland waters. These included: (1) water management to be performed with the limits of river basins; (2) the necessity of considering jointly problems of water supply and pollution control; and (3) administrative units concerned with water management to prepare a regional water plan. With these principles for guidance, the intent is to replace the existing water management districts with ten water management regions. With the exception of some highly industrialized areas, the ten regions will be based on catchment basins.

In 1988 a draft of the 'National Programme for Natural Environment Protection to the Year 2010' (National Programme . . . , 1988) was developed by the Ministry of Protection of Environment and Natural Resources in collaboration with its Institute of Environmental Protection. The draft was subjected to wide discussion in scientific, economic and

Figure 7.3 Map of Poland showing the *vojevoidship* (provinces) and the division of Water Management Directorates (after General conception . . . , 1987)

social circles. The development of the draft resulted from the implementation of recommendations of congresses, political parties and the Patriotic Movement for National Restoration. The latter is the social organization which was created after the cancellation of martial law in Poland during 1981/82. Its purpose is to promote the consolidation of the nation.

In January 1985, the President of the Government of the Polish People's Republic had decided to initiate the preparation of the 'Programme', and in December 1985 the Cabinet had ratified the concept of the Programme. Finally, in September 1986, the Prime Minister had established a special team to prepare a draft of the 'Programme' under the leadership of the Minister of Protection of Environment and Natural Resources. As a result, the Programme has had the status of an official political and government document, which should ensure that it will be implemented.

The 'National Programme for Natural Environment Protection to the Year 2010' has the following characteristics: (1) *complex*, in that the impact of economic development on the environment must be considered and in that environmental and ecological conditions which can support economic and social development must be examined; (2) *interdisciplinary*, in that human, economic and organizational as well as technical and technological aspects must be considered regarding environmental protection; (3) *spatial*, in that attention must be given to strategies to eliminate or minimize threats to the environment at scales ranging from the local community to the nation; and (4) *open*, in that modifications to plans are to be expected based upon discussion with interested and affected people and groups, consideration of research results and consideration against other economic and social plans.

The Programme is designed to check and reverse unfavourable tendencies of environmental pollution and degradation, to restore the environment to standards stipulated by the law, to avoid water shortages and ecological threats and thereby to ensure proper functioning of natural ecosystems, required economic growth and suitable social living conditions. Against those purposes, the key resolutions concerning water management in the Programme can be examined.

The ecological situation in Poland was analysed in the context of Europe as a whole. Twenty-nine areas comprising 11 per cent of the area of Poland were judged to be under ecological threat, in that pollution standards have been exceeded. These 29 areas are inhabited by 34 per cent of the population of Poland and represent the most industrialized and urbanized parts of the country. Numerous sources were found to be impairing water quality of surface waters, including municipal and industrial wastes, saline mine waters primarily from coal mines, surface runoff from fertilized arable land, urban, industrial areas and roads and air pollution.

Although some 200 industrial and municipal sewage treatment plants are put into operation every year, the quality of surface waters has not been improving. Only slightly over four per cent of the length of the rivers investigated in 1986 had class I water quality, a level of purity suitable for drinking and for sustaining salmon. Class II and III waters each covered 28 to 29 per cent of the nation's rivers, while about 40 per cent of the rivers did not meet acceptable standards. Class II water is suitable for livestock watering, other forms of fish and recreation, while Class III is adequate for industry and irrigation. Groundwater, used mainly for municipal purposes and by the food processing industry, is also suffering from increasing pollution.

The possibilities of compensating for the extremely variable water runoff, the main water resource of Poland, are minimal. The total capacity of storage reservoirs together with dammed lakes is equal to just 6 per cent of the average annual river runoff. Meanwhile, the current demand is estimated to be at 10 to 12 per cent of the average annual runoff and there are not any good sites for additional dams and reservoirs.

The largest water consumption (70 per cent of water consumption occurs in eight of the 49 provinces) is in regions with limited sources of water supply, which is a great drawback for future economic growth. Water supply for the population is not satisfactory. About 25 per cent of the cities have constant or periodic water deficits and less than 50 per cent of the people living in the country have water supply systems. The irrigation of arable land is still poorly developed. The Programme shows that the degradation of the environment is slowly becoming a barrier to economic development.

The foundations of the Programme were based on anticipated social and economic development, including improvement in the quality of life for the Polish people and adaptation of the economy for the needs of the twenty-first century. The population of Poland was projected to increase to 40 million people by the year 2000. It is planned that economic development will lead to self-sufficiency in food production and provide satisfactory nutrition for the entire population.

The report 'Dilemmas of the Development of Poland on the Threshold of the 21st Century', prepared by the Committee for the Prognosis of the Country's Development 'Poland 2000' of the Polish Academy of Sciences, was used. The main ecological goals to the year 2010 were identified. With regard to water management, these goals included: (1) elimination of water pollution in rivers and lakes currently not meeting standards, (2) restraint of groundwater degradation before the year 2000, and (3) significant improvement of the quality of nearshore seawater. These goals are conditional upon the changes to the structure of the national economy already under way. These changes include the closing of plants that generate excessive pollution, the use of and compliance with regulations and standards and suitable control systems.

The Programme identifies a spatial planning framework for the country by specifying areas of ecological threat, areas of protected landscape (protected against urbanization and industrialization due to specific landscape features and values) and areas of ecological equilibrium (protected against all human activity to serve as biosphere reserves). The Programme also identifies appropriate technological policies and organizational activity which encourages scientific research and construction of the instruments and equipment for environmental protection. The Programme considers the international obligations of Poland resulting from signed conventions, declarations and international treaties concerning the protection of border waters and the Baltic Sea.

Part 2 of the Programme is concerned with strategies for the protection of the natural environment of Poland. Within the domain of water management, and of water supply in particular, the recommendations are for: further intensive construction of sewage treatment plants, building of multiple purpose storage reservoirs and smaller reservoirs and transport of water and wastes. These activities should help to restore the capability of Polish rivers to achieve self-purification. It also is intended to rationalize water consumption, and to introduce policies for industrial location which reflect accessibility to water resources and which encourage water-saving and waste-minimizing technologies. Finally, the Programme intends to introduce integrated water supply and sewage disposal systems and to upgrade existing systems.

In order to implement these plans, adequate funds have to be made available. The Programme has provided for an increase in discretionary funds, utilization of societal funds for ecological expenses and significant participation by representatives of the local communities, primarily local representatives from people's councils of every rank, leagues, clubs, associations, committees and even so-called informal organizations such as the 'green movement'. The broad-based ecological education in Poland, both formal and informal, favours the realization of incorporating the local communities into the process.

Conclusions

Poland belongs to the group of countries which do not have very abundant water resources, particularly because of a growing population. Not only is the quantity of water resources unfavourable, the spatial and temporal distribution over the country is unfavourable. As early as the 1950s, these problems led to a perspective plan for water management which provided the main ideas for a long-term approach. Nevertheless, as industry and urbanization were growing rapidly, considerable water pollution problems emerged. In that difficult situation it was found useful, at the beginning of the 1960s, to create an office of water management, first as a central board and then as a ministry, to co-ordinate and invest when large hydrotechnical multipurpose structures were built.

For managing Polish water resources, great importance has been given to establishing a water law which conforms to a socialist national economy. Such a law was introduced in 1962 and then modified in 1974. As regions with water shortages emerged, attention was given to modifying regional water administration. Several territorial boards of water management were created to preserve and develop open waters (rivers, lakes, canals). The boundaries of such boards were not the same as those for catchment basins.

In the mid 1970s, the regional administration in Poland was recognized. Forty-nine provinces were formed. Local water problems and the granting of water law licences were allocated to provinces. The emerging water crisis, regarding both water quantity and quality, particularly acute in some regions combined with the potential threat of water shortages for future economic development, require new approaches and action. Particular attention needs to be given to resolving water pollution, and there is a great need to build more sewage treatment plants. Water demands will require the construction of reservoirs to increase the water retention capacity in the country. An overriding need is to improve the co-ordination and integration of various aspects of water management, and water management with other related resources and societal concerns. An important step forward will be the creation of water management regions which are based on catchment basins rather than administrative units.

Notes

1. According to the *Statistical Yearbook*, 1988, Poland had 34.2 million inhabitants in 1975. The population in 1950 was 25 million.
2. As a result of war damage and wasteful exploitation by the invaders at that time, the forests in 1946 covered only 21 per cent of Poland. This increased to about 28 per cent in 1987 (*Statistical Yearbook*, 1988).
3. Unfortunately, only one barrage was built here then. Additional ones were to be built at the end of the twentieth and beginning of the twenty-first centuries.

References

'General conception of the organizational model of water management in Poland', 1987, *Gospodarka Wodna*, **1** (47), pp. 2–5 (in Polish).

Gutry-Korycka, M., 1985, 'The structure of natural water balance in Poland 1931–1960', *Prace i Studia Geograficzne*, **7**, pp. 91–134 (in Polish).

L'vovich, M. I., 1979, *World Water Resources and their Future*, American Geophysical Union, Washington, DC.

Mikulski, Z., 1982, 'State of scientific research on water management in Poland', *Acta Academiae Scientiarum Polonae*, **3–4** (1), pp. 65–77.

'National Programme for Natural Environment Protection to the Year 2010' (draft), 1988, Ministerstwo Ochrony Srodowiska i Zasobow Naturalnych, Warsaw (in Polish).

Statistical Yearbook, 1988, 1988, Glowny Urzad Statystyczny, Warsaw (in Polish).

Water Law, executive and union regulations, 1979, Wydawnictwo Prawnicze, Warsaw (in Polish).

CHAPTER 8

Integrated water management: the Nigerian experience

Ademola T. Salau

Introduction

Water is one of the most fundamental of natural resources that a country must harness in its quest for a rapid rate of economic growth. The role of water in the development process cannot be overemphasized, particularly for a developing country such as Nigeria. The demand for water has increased tremendously over the years in Nigeria and will continue to increase in view of the accelerating tempo of urbanization, population growth and industrialization.

Efficient management of water resources is important to a country in which the overwhelming majority of the population still depends on agriculture for its livelihood. Even in a country such as Nigeria with abundant rainfall, water constitutes a major constraint for agriculture in terms of the number of crops that can be grown in a year. Water resource development is also significant for improvement of the living standard of the people, particularly the rural dwellers. Rural water development is important for encouraging investors to locate in rural areas so as to create more jobs.

Thus, it is not surprising that successive administrations in Nigeria have placed a high priority on the provision of water for domestic, agricultural and industrial use. Government expenditure for the provision of water supplies has increased tremendously in absolute terms over the years (see Table 8.1).

In spite of the large amount of money invested in water resources over the years, the provision of water has remained a major problem. In view of the failure of different approaches to solve this problem, and with the fear of an imminent 'food crisis' in the late 1970s, a river basin development strategy was adopted. This river basin development strategy was seen as utilizing an integrated water management approach capable of achieving optimum use of natural water resources on a basin-wide scale. The question is how well this new approach has worked in practice? Has it

Table 8.1 Financial allocation to water supply
provision

Plan period	Capital expenditure on water (million)
1946–56	17.124
1955–60	35.000
1962–68	54.000
1970–75	103.400
1975–80	930.039
1981–85	5,371.653

Source: Adapted from Ayoade and Oyebande (1978)

really provided a framework for the integrated development of Nigeria's
water resources? If not, what are the factors responsible for this? These are
the issues addressed in this chapter.

Land and water resources of Nigeria

Land

The geographical area of Nigeria is 923,768 km^2 (98.3 million hectares) and
it is situated within longitude 2°36′–14°35′E and latitude 4°–14°N. The
temperature is fairly constant throughout the year being between 20° and
35°C during the day. Annual precipitation varies from north to south. The
country can be conveniently divided into three ecological zones:

1. the humid forest zone covering about one-fifth of the land area;
2. the humid–sub-humid savanna zone occupying about three-fifths of the
   land area; and
3. the semi-arid zone consisting of the Sudan and Sahel Savanna which
   occupies approximately one-fifth of the land area.

   Of the total land area, approximately 71.2 million hectares are judged to
be cultivatable. However, only one-half of this amount is under actual
cultivation (Nigeria, 1971, p. 98). Table 8.2 shows some aspects of land use
for 1951 and 1976 and the projection for 1986. It can be seen from this table
that land use in the country has undergone rapid changes. Land under farm
and tree crops, which covers roughly the two major productive sectors of
agriculture, used 10.6 per cent of the total land area in 1951. By 1976, this
category of land use had increased to 18 per cent and it was estimated that

Table 8.2   Land utilization in Nigeria for the years 1951, 1976 and 1986

Category of land	1961 %	1976 %	1986 %
Uncultivated range land primarily used for livestock	67.0	50.0	39.0
Fallow farmlands 40% of which is usable for grazing	13.8	17.0	20.0
Non-agricultural land, including towns, roads and airports etc.	1.0	5.0	7.0
Land under farm crops	9.4	15.0	20.0
Land under tree crops	1.2	3.0	4.0
Land under forest reserves 33% of which is usable for grazing mainly in the North	7.6	10.0	10.0
Total	100.0	100.0	100.0

Source: National Council for Agriculture Memorandum (1980) Lagos (cited by M. O. Awogbade, 1980)

it would increase further to 24 per cent of the total land area by 1986. Thus, the amount of land under cultivation was still very low, representing approximately a quarter of the total land area in 1986.

In 1951, 13.8 per cent of the land was left in fallow and this increased to 17 per cent in 1976. Land classified as 'uncultivated' range land constituted about 67 per cent in 1951 but decreased to 50 per cent in 1976 and it is estimated it will decrease to 39 per cent by 1986. A portion of this category of land consists of mountainous or hilly and eroded land. However, a part of this is also arable land left uncultivated which is being brought under cultivation as the need arises. Since this land is also required for livestock grazing, the implication is that this is not 'idle' or 'vacant' land in the real sense.

Many factors are responsible for the ineffective utilization of the nation's land. Primary reasons relate to the physical conditions, featuring excessive rainfall in some areas leading to extensive flooding, and poor drainage and inadequate moisture in other areas. These factors have combined to restrict the area of land suitable for cultivation.

Water resources

Rainfall.   The pattern of agriculture and the length of the growing season are dependent on the distribution of annual rainfall. According to Kowal and Knabbe (1972), the dominant features of rainfall in Nigeria are its seasonality, large energy content and variability from year to year. The energy content of rainfall is high and the amount of water held and the rates at which it is shed are also large. The result is serious flood problems and sheet and gully erosion in some areas.

The mean annual rainfall varies from more than 3500 mm in the south to about 1000 to 1500 mm in the middle belt, and dropping to less than 500 mm in the semi-arid parts of the north-west (Agboola, 1979). Thus, while a water surplus usually exists on an average annual basis in the south, there is considerable water shortage in the northern part of the country. Perhaps more important, however, is the fact that every part of the country has a period of soil moisture deficiency varying from a few weeks in the south to many months in the north (Are and Fatokun, 1984). Hence, control of water, water management by irrigation, drainage and management of the land and soil surface are of extreme importance in the country.

Surface water. Nigeria is endowed with many river systems and basins. Major ones include the Niger, Benue, Gongola, Hadeijia, Ogun, Imo, Anambra and Cross Rivers. These rivers and streams, however, are unevenly distributed over the country and most of them, especially in the north, carry less water and easily dry up during the dry season. Many of these rivers also suffer much loss of water through evapotranspiration and infiltration (Ayoade and Oyebande, 1978). None the less, in spite of these problems it is believed that the country still has sufficient surface water resources which, if properly managed, could meet the immediate identifiable needs.

Groundwater. About 60 per cent of the country is underlain by crystalline rocks of the basement complex which are generally known to be poor aquifers. In general, these rocks have no primary permeability. Groundwater therefore occurs in fractures, joints and partially decomposed rock. In comparison to the sedimentary rock with which the rest of the country is underlain, and which yields 30,000 litres per hour of water, the basement complex has a water yield of about 4,000 litres per hour (Iwugo, 1986). Aquifers have been located between 20 and 80 metres depth, with an average depth of 50 metres. Groundwater production from some boreholes in the Nupe sandstone of the basement complex has been found from depths of up to 250 metres (Beale, 1988, p. 102). In the drier parts of the country, water supply relies to a great extent on the groundwater sources.

Water resource planning and management

As in most African countries, until very recently virtually all efforts in Nigeria in water resource planning and management were directed towards satisfying domestic and industrial needs in general, and relieving shortages which often arise in the urban centers (Oyebande, 1975). Clean and adequate supplies of water to the burgeoning urban centres have constituted a critical problem to successive administrations in Nigeria. From the colonial period onwards, water resource planning and development had been the main preserve of the State Water Boards and Corporations. Traditionally, drinking water is usually procured from hand-dug wells or nearby streams and thus a high incidence of debilitating water-borne diseases occurs. The first modern water supply scheme was completed in Lagos in 1914. In 1953, there were twenty-seven water supply schemes in the country. These increased to sixty-seven by 1960 when the country attained independence from Great Britain. Although since this time almost all the major towns have been provided with access to pipe-borne water, the supply of clean and adequate water is still a major problem throughout the country.

An average of 122 litres per person per day of water has been estimated as the ideal minimum requirement for all purposes. Most Nigerian cities, however, are served with well below this minimum requirement with only two states (Kaduna and Niger) achieving this target. More important, perhaps, is the fact that eight states received less than half the requirement in 1977 (Oyebande, 1978). The frequent disruptions of water supply services are testimony to the ineffective performance of the various water boards and corporations.

The ineffective performance of these organizations to plan and develop the nation's water resources is due to many factors. One of the most important is the lack of adequate and accurate information on the quality and quantity of the nation's water resources. There is also the problem of the absence of accurate information about the population of consumers to be served and their water needs. The Federal Government, however, has contended that 'the deficiencies in the water supply situation are largely a development problem which can be remedied by improved planning efforts' (Nigeria, 1975, p. 297). However, there has been a failure by the Government to identify what these 'improved planning efforts' would involve. Nevertheless, the fact is that public water boards and agencies usually have fragmented and shared responsibilities, cutting vertically through different levels of governments (local, state, federal) and horizontally through different ministries or agencies (water, agriculture, works, wildlife and forestry etc.). The result is duplication and lack of co-ordination and thus, ineffective water management.

Towards comprehensive water resources development and management

In view of the need to achieve and maintain a rapid rate of economic growth, and to diversify the nation's economic base from the excessive reliance on one commodity (oil), the government from the late 1970s onwards has become more and more conscious of the need to appraise the nation's resources. Resource appraisals became necessary, especially in view of the economic problems facing the country. It was realized that, although lack of capital and social institutions constitutes a hindrance to agricultural development, factors such as climate and water also pose serious impediments. For example, climatic and water constraints, in the form of insufficient and unreliable rainfall, especially in the northern part of the country, were recognized as a hindrance to the implementation of agricultural programmes. In the First National Development Plan (1962, p. ll), the importance of the need to take stock of water resources was found to be an essential prerequisite for large-scale planning in agriculture and forestry resource development, especially in those requiring irrigation works. It was recommended at this time that a 'nation-wide survey' of Nigeria's water resources needed to be conducted.

In the Second National Development Plan (1970–4), the case for the exploitation and harnessing of the country's water resources for the purpose of agricultural development was made on the grounds that

in more than three quarters of the northern part of the country, rainfall is not more than five months duration. In addition, the beginning of the wet season is quite variable from year to year. Thus, full exploitation of the agricultural potential of the country is only possible if its water resources are developed for irrigation, which will not only permit double cropping, but will also ensure sufficient water during the growing season of the crops (Nigeria, 1970, p. 110).

The Study Group on Irrigation and Drainage of the National Agricultural Development Committee, which was created in 1971 to examine the problem, presented a report strongly arguing that 'available land and water resources in Nigeria, if developed scientifically under irrigation and drainage, shall perhaps contribute the largest single factor for increasing agricultural production. The unused potential in this sector is of considerable magnitude' (Nigeria, 1971, p. 5). The report also examined the possible consequences of not instituting irrigation projects, particularly in the northern part of the country. Furthermore, even for the southern part of the country, which generally has sufficient rainfall, the Study Group pointed out that there would be increased agricultural productivity through the development of irrigation because:

in much of the southern states irrigation is needed to permit double cropping or to increase yields of plantation crops. In these areas, considerable amounts of natural resources is being wasted in flooded and hill drained lands lying along the river banks. These areas if reclaimed and provided with adequate water control and drainage can be highly productive for growing crops like fibre and rice.

This line of thought added 'to the impact of four or more successive years of droughts in the Sahelian and sub-Sahelian zones of Nigeria'

(Taylor, 1981, p. 11). Subsequently, in the early seventies, Nigeria had to make 'irrigation development become a *sine qua non* for a dependable, viable agriculture' (Liman, 1981, p. 5).

Although river gauging in Nigeria dates back to 1908 on the river Niger (Nwa, 1978), the idea of river basin management was first mooted by the Food and Agricultural Organization (FAO) in a report submitted to the Federal Government in the early 1960s. The tentative implementation of the FAO report resulted in the formation of the Niger Delta Development Board in 1960. In the First National Development Plan (1962–1968), the Lake Chad basin, River Niger Basin and River Niger Commission were formed. In the Second National Development Plan (1970–1974), the Kainji Lake Development Commission and the Sokoto River Basin Development Authorities were established. The establishment of the Federal Ministry of Water Resources was the first sign of the serious attention to be given to water resources in the Third National Development Plan (1975–1980). The culmination of this was the creation of eleven River Basin Development Authorities (RBDA) by the Federal Government through the promulgation of decrees. The country was, in effect, divided into eleven river basin regions (see Figure 8.1).

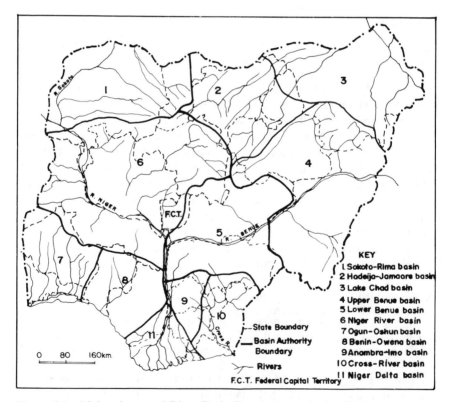

KEY
1. Sokoto-Rima basin
2. Hadeija-Jamaare basin
3. Lake Chad basin
4. Upper Benue basin
5. Lower Benue basin
6. Niger River basin
7. Ogun-Oshun basin
8. Benin-Owena basin
9. Anambra-Imo basin
10. Cross-River basin
11. Niger Delta basin

-- State Boundary
— Basin Authority Boundary
〜 Rivers
F.C.T. Federal Capital Territory

Figure 8.1 Major rivers and River Basin Development Authorities' areas

Decree number 87 of 1979, Section Four, identified the functions of the RBDAs as follows (Salau, 1986):

1. to undertake comprehensive development of both surface and underground water resources for multipurpose use;
2. to undertake schemes for the control of floods and erosion, and for watershed management;
3. to construct and maintain dams, dykes, polders, wells, boreholes, irrigation and drainage systems and other works necessary for the achievement of the authority's functions;
4. to develop irrigation schemes for the production of crops and livestock and to lease the irrigated land to farmers or recognized associations in the locality of the area concerned, for a fee to be determined by the authority with the approval of the commissioner;
5. to provide water from reservoirs, wells and boreholes under the control of the authority for urban and rural water supply schemes on request by the state government and when directed to do so by the commissioner;
6. to develop fisheries and improve navigation on the rivers, lakes, reservoirs and lagoons in the authority's area;
7. to process crops and livestock produced in the authority's area;
8. to control pollution in rivers and lakes in the authority's area in accordance with nationally laid down standards;
9. to resettle persons affected by the works and schemes specified in paragraphs 4 and 5 above or under special resettlement schemes.

The authorities are autonomous but their activities are supposed to be co-ordinated by a relevant government ministry.

River basin development as integrated water management

The establishment of River Basin Development Authorities represented a significant shift in approach regarding the management of water resources in Nigeria. According to Saha (1981), river basin planning seems to offer a comprehensive framework for the integrated development of the environment which can bypass inadequate agencies and procedures. The major objective of river basin planning is to achieve the fullest possible use of all resources in the basin with water serving an integrative role. The key strategies of river basin development are the construction of dams, canals and levees for purposes of irrigation, navigation and hydroelectric power; flood plain development schemes to reclaim flood-prone areas; and schemes for soil conservation, recreation and afforestation.

The river basin approach has been adopted in Nigeria in the hope that the resulting integrated control of the nation's water resources would improve water supplies for agricultural, industrial and domestic purposes. However, as can be seen from the list of dams already built, under

Table 8.3   Examples of dams already built, under construction or proposed for the future

Name	Project description and status	Estimated cost
Nyaba Project (Anambra State)	Comprises two dams on Nyada river to bring about 11,500 ha under irrigated agriculture for crops and livestock production	5.300m
Upper Imo River Project (Imo State)	A dam to be built across Imo River to provide irrigation water for 3,500 ha at Ndimoko. A dam to be built also at Ibu to provide drinking water	22.200m
Ose-Ossiomo River sub-basin projects	A major project to supply water for multipurpose uses	20.286m
Dadin Kowa Project	An ongoing project involving the construction of a rockfill dam spillway, diversion conduit, irrigation release facility etc. on Gongoloa River for multipurpose development–irrigation, h.e.p., fishery, domestic and industrial water supply (*An integrated strategy*)	72.00m
Begel-Zungur Irrigation Project	Project envisages the construction of earth fill dam across Begel River for multipurpose uses—3,000 ha of irrigated agric, livestock fattening, dairy production, pasture development and rural and urban water supply	1.00m
Upper Benue Valley Project	Dasin Hansa multipurpose project which is expected to involve a dam of 20,000 million cubic metres capacity, waterway development	2.00m
Suntai River Project	A plan to harness the Suntair River for multipurpose development including irrigated agriculture over some 12,000 ha for sugar cane cultivation and hydropower generation of about 200 MW capacity	2.119m
Gubi Dam Project	A multipurpose project will be constructed at Bubi for irrigation of 200 ha and for urban water supply	5.0m
Mada River Project	An ongoing project involving the construction of 2 dams at Doma and Tede for power development and rice cultivation	25.0m
Shemanker River Basin Project	An ongoing project involving the construction of 4 dams within the Shemanker River basin to provide all year round water for irrigated agriculture (Dansack Dam, Shendam Dam, Bakshe and Shemanker Dams)	10.0m
Alau Dam and Jere Bowl Rice Scheme	Project envisages construction of a dam on River Ngadda to store water which will be used both for urban consumption and for irrigation of about 8,000 ha in the Jere Bowl for rice and vegetable production	3.5m

Table 8.3 *contd*

Name	Project description and status	Estimated cost
Aya River Project	A dam will be constructed on River Aya to provide irrigation water for cultivationof lands of about 1500 ha	8.820m
Idim Ibom/Nkeri Project	Project involves the construction of two dams to impound water for irrigating 19,000 ha of land	8.640m
Challawa Gorge Dam	An ongoing project for the construction of a dam for h.e.p. and fisheries	20.00m
Jama'are Valley Project	Involves the construction of Gilala Dam and completion of Kafin-Zaki Dam and Kawali diversion structures for irrigated agriculture, livestock and fisheries development	34.560m
Kainji Lake Basin Project	This involves the construction of a concrete dam at Kubli, a diversion dam on Swashi River and 30 km canal across Kontagora River for integrated agriculture and fisheries schemes	28.00m
Oyan Dam and Resettlement Scheme	An ongoing project involving the construction of an earthfill dam on Oyan River for irrigated agriculture and h.e.p., rural and urban water supply and fishery and supply production *(An integrated strategy)*	65.560m
Ikere Gorge Dam	Involves the construction of earth dam for the provision of water-irrigated agriculture, fishery and livestock development and portable water for some rural and urban settlements	55.960m
Bakalori Project	An ongoing project involving construction of concrete and earth dams and canals for irrigating 23,200 ha of land and 3,000 KW h.e.p., flood control, livestock and fish production *(An integrated strategy)*	219 m
Goronye Dam Project	Involves the construction of three dams totalling 12,530 m designed to store and supply water for irrigated cultivation of about 32,000 ha and h.e.p. station of 2,000 KW capacity	264.500m
Karaduwa River Basin Project Zobe Dam	Involves the development of year-round cultivation of an area of about 11,770 ha by water supplied from Karaduwa and Zobe Dams to be constructed under the project	99.500m

Source: Nigeria, 1981, *Fourth National Development Plan, 1981–85*, vol. II (The National Planning Office, Lagos)

constructon or proposed for the future, most of these schemes fall far short of this aim (see Table 8.3). Apart from the fact that most of the schemes were not part of an integrated model, the following are also factors which inhibit the integrated management of the nation's water resources by the River Basin Authorities.

Management

In spite of the general acceptance of the concept of integrated water management through the establishment of RBDAs, actual implementation and day-to-day operation have been far from integrated in nature. The various basin authorities have directed their efforts mainly towards dam construction for water supply and irrigation projects. For example, by September 1984, the Tiga Dam and the Kano River project was able to irrigate 27,750 hectares out of a 134,000 hectares target, while the Bakalori project had been able to meet the planned target of 25,000 hectares. It was also reported that twelve of the then existing eighteen RBDAs assisted their participating farmers to plant 188,194 hectares of various crops during the 1984 planting season (Are and Fatokun, 1984).

There has been little work done in the areas of watershed management, afforestation, pollution control, navigation, fisheries and even resettlement as mandated by the decrees setting up the RBDAs. The resettlement record of the basin authorities is particularly poor and has constituted a serious impediment to the realization of irrigating the many thousand hectares of land as planned. For example, the Tiga Dam in the Hadeijia-Jamaare Basin displaced 13,000 farmers and villagers who were supposed to be resettled in seven new villages around the project area. Many of these people refused to move and those who did move encountered numerous problems such as poor quality of farmland and infrastructures in the area. Kiri Dam in the Gongola Basin displaced over 20,000 people and almost the same number of people were displaced by the Bakalori Dam in the Sokoto-Rima Basin. The scale of expenditure required to carry out re-settlement and pay compensation to the affected people usually exceeded the basin authority's financial resources. The result was that resettlement, if and when carried out, was usually shoddily done with compensation remaining unpaid for many years. Thus, animosity of the local people continued to rise and culminated, as in the case of the Bakalori project, in community protests which were stopped at the cost of many lives.

Further, there have not been any deliberate efforts to integrate the separate basins. Adams' (1985, p. 301) contention that the RBDAs represent a facade of integrated planning aptly summarizes the situation. As Adams pointed out, in a number of river basins the development of particular dams or irrigation projects preceded the establishment of the RBDAs. In many cases, project development preceded integrated appraisal of water resources available in the basins. For example, it was not until 1977 that a study was commissioned by the Federal Ministry of Water Resources to investigate the long-term development possibilities of the Gongola Basin. By the time the study was concluded, contracts for the construction of Kiri Dam had already been awarded, in spite of the study's conclusions that the proposed dam was too big and that development should start with flood control dams in the headwaters of the basin.

Administration framework

With a three-tier system of government (local, state and federal), the responsibility for water supply is legally vested in the state and local government authority, while the Federal Government co-ordinates all activities related to water resources development and management. Nevertheless, there is an overlap in the functions of the various bodies dealing with one or other aspects of water resources established by the three levels of government. The states often pursue different water development policies through their water boards and water corporations, leading to excessive bureaucracy and inefficiency in planning. For example, although the Ogun–Oshun River Basin Development Authority (OORBDA) operational area covers Oyo, Lagos and most parts of Ogun State, it shares responsibility for water planning in Oyo State with the Oyo State Water Corporation (OSWC), which was entrusted with day-to-day management of water supply in the state. The result was that OORBDA and OSWC have overlapping and duplicating functions.

Perhaps more important is the fact that the nation's politico-administrative system does not encourage integrated management of water resources. For example, although the River Basin Development Authorities were established in the hope of attaining integrated development of the country's water resources, in practice they did not receive much co-operation from the state governments, especially of those governed by political parties different from that at the federal level. The result was that co-ordination and effective management became difficult in such situations. In addition, there was also the problem of lack of proper delineation of the area of influence of the RBDAs. Many were actually defined closely following state boundaries rather than on physiographic terms in spite of the fact that river basins cut across the state boundaries.

Operational constraints

In spite of the general acceptance of the concept of integrated water management by the River Basin Development Authorities, actual implementation is bedevilled by many problems. There is insufficient information available to assist planning and management decisions. Most of the basin authorities had no hydrological/meteorological data and some, such as the Benin–Owena River Basin Development Authority, have yet to establish a hydrometeorological station.

The basin authorities had inadequate financial resources to execute their projects. Thus some 'projects which took years of painful feasibility studies, aerial photography and detailed engineering efforts to design like the Lower and Middle Ogun project of the Ogun–Oshun River Basin Authority cannot take off' (Are and Fatokun, 1984). The authorities were also faced with a shortage of qualified and experienced technical manpower and thus the design and planning of some of their projects had to be contracted out to consultants and foreign companies.

The River Basin Development Authorities have also not been able to reduce the cumbersome bureaucracy and inefficiency in water resource development and planning in Nigeria. There is at present some overlap in the composition, areas of jurisdiction and functions of the River Basin Authorities, various state water corporations and the national water resource development committees. For example, there was conflict between the Ogun–Oshun River Basin Development Authority and the Oyo State government when the former attempted to take over Idere Gorge Dam from the state government on the grounds that it should come under its sphere of influence. The state government resisted and won.

Conclusion

At present, the practice of river basin planning in Nigeria falls far short of what can be described as integrated or comprehensive water resource management. The River Basin Development Authorities have focused on dam construction and irrigation development. Little attention has been paid to other areas such as watershed management, fisheries, navigation and pollution control. Perhaps more important is the fact that river basin planning has failed to streamline the existing cumbersome administrative framework and thus avoid the friction and lack of co-ordination bedevilling water resource management in the country.

Land use planning and water resources development need to be integrated. The fact is that all of the activities within a basin need to be co-ordinated in relation to other resources of the basin. The intensity and type of land use within the watershed must be studied and planned for. The impact on soil of sudden and significant increases in water to large areas of land can be detrimental. For example, it is well known that the porosity of some Nigerian soils is low and thus a large application of water on them may lead to high water runoff and a tendency for the soil to compact under wet conditions. River Basin Development Authorities provide a suitable administrative structure for implementing water resources management functions on the national and state levels, and thus their competition with various state boards and corporations is really unnecessary.

Policy priorities for the future in water resources development must include the need to solve the issue of jurisdiction between the river basin authorities and the state apparatus. One way to do this is to entrust the river basin authorities with wider responsibilities and power relative to the state corporations.

In order for the basin authorities to attract the highly technical manpower they require, the remunerations paid should be more attractive than those being paid to civil servants.

There is also need for continuity in policies by different administrations at the federal level. This is probably better done by providing for the existence of these authorities in the nation's constitution. Hence, one cannot but conclude that integrated water resource development in Nigeria is still a goal towards which to strive.

References

Adams, W. M., 1985, 'River basin planning in Nigeria', *Applied Geography*, **5**, pp. 297–308.

Agboola, S. A., 1979, *Agricultural Atlas of Nigeria*, Oxford University Press, London.

Are, L. and Fatokun, J., 1984, 'River Basin Development Authorities and irrigation', paper presented at the Conference on Strategies for the Fifth National Development Plan 1986–1990, organized by NISER and the Federal Ministry of National Planning at Ibadan, 25–9 November.

Are, L., Fatokun, J. and Togun, S., 1982, 'Experiences in river basin development with special reference to the Ogun–Oshun River Basin Development Authority', Proceedings of the 4th Afro-Asian Regional Conference of the International Commission on Irrigation and Drainage, Lagos, Nigeria, vol. 1, pp. 99–110.

Awogbade, M. O., 1980, 'Livestock development and rangeland utilization in Nigeria: the need for an integrative approach', paper presented at the Conference on the Commission on Nomadic Peoples, Nairobi, Kenya, 4–8 August.

Ayoade, J. O. and Oyebande, B. L., 1978, 'Water resources' in J. S. Oguntoyubo *et al.* (eds.), *Geography of Nigerian Development*, Heinemann, Ibadan, pp. 71–88.

Beale, G., 1988, 'Improved groundwater supplies for Central Nigeria' in *Developing World Water*, W.E.D.C. and Loughborough University, Grosvenor Press.

Iwugo, K., 1986, 'Groundwater quality treatment and pollution in Nigeria—Lagos Metro case study', paper presented at the 1st annual Symposium on Groundwater in Nigeria.

Kowal, J. M. and Knabbe, D. T., 1972, *An Agroclimatological Atlas of Northern States of Nigeria*, A.B.U. Press, Zaria.

Liman, M., 1981. 'Overview of the crop subsector', in F. S. Inachaba (ed.), Proceedings of a Workshop on the Crop Subsector in the Fourth National Development Plan 1981–5, Federal Ministry of Agriculture, Lagos.

Nigeria, 1970, *Second National Development Plan, 1970–1974*, Federal Ministry of Planning, Lagos.

Nigeria, 1971, *National Agricultural Development: Committee Report of the Study Group on Irrigation and Drainage*, Federal Ministry of Agriculture, Lagos.

Nigeria, 1975, *Third National Development Plan, 1975–80*, Federal Ministry of Planning, Lagos.

Nwa, E. U., 1978, 'Water resource development, utilization, and management for agricultural production in Nigeria', *Sumaru Conference*, paper 16, Institute of Agricultural Research, Zaria.

Oyebande, L., 1975, 'Water problems in Africa' in P. Ricardo (ed.), *African Environment: Problems and Perspectives*, I.A.I., London.

Saha, S. K., 1981, 'River basin planning as a field of study: design of a course structure for practitioners' in S. K. Saha and C. J. Barrow (eds.), *River Basin Planning: Theory and Practice*, John Wiley, London, pp. 10–40.

Salau, A. T., 1986, 'River basin planning as a strategy for rural development in Nigeria, *Journal of Rural Studies*, **32** (4), pp. 321–35.

Taylor, T. A., 1981, 'A review of the crop subsector in the Third National Development Plan' in F. S. Inachaba (ed.), Proceedings of the Workshop on the Crop Subsector in the 4th National Development Plan, 1981–5, Federal Ministry of Agriculture, Lagos.

CHAPTER 9
Patterns and implications
Bruce Mitchell

1. Introduction

In the first chapter, it was argued that integrated water management may be interpreted in at least three ways: integration of different components of water, integration of water with related land and environmental resources and integration of water with social and economic development. The experience from the seven countries considered in this book suggests that the motivation for an integrated approach has most often started at the second level, in which concern is directed to the linkages among water, land and environmental resources.

Early interest usually centred upon efforts to address problems concerning erosion and other forms of land degradation. This aspect was certainly central in New Zealand with the passage of the Soil Conservation and Rivers Control Act 1941, and in Canada with the creation of the Ontario Conservation Authorities in 1946. In other instances, a wider range of concerns was integrated. Thus, in the United States, the Tennessee Valley Authority emphasized control of flood damages, generation of hydro-electricity power and improvement of river-based transportation, while in England multiple-function river basin agencies were established.

There still seems to be fragmentation of water components among management agencies, and it is not uncommon for surface and ground-water to be handled by separate agencies, and for clean and dirty water to be dealt with by different organizations. In that regard, integration seems further advanced in many countries regarding the interrelationships among water, land and environmental resources than it does in linking the different components of surface and groundwater or quantity and quality of water.

A few countries have sought explicitly to integrate water with social and economic development. In Japan, effort has been made to relate water resources management to National Comprehensive Development Plans. In Poland, national economic plans set the context for water planning, while in Nigeria a series of National Development Plans set the priorities for water management. While it is helpful to integrate various aspects of water with land and other environmental resources, in the long run the goal of integrated water management should be to link water management to social and economic development. For most of the experiences discussed in this book, that long-run goal is still to be realized.

In the opening chapter, a distinction was also made between *normative*, *strategic* and *operational* levels of management. It will be recalled that at a normative level attention is focused upon what ought to be done. At a strategic level interest centres upon what can be done, whereas at an operational level the focus is upon what will be done. There seems to be growing consensus at the normative level that it is desirable to consider the interrelationships among the components of water, the linkages among water, land and environmental resources, and the ties between water and social and economic development. The latter aspect has been particularly emphasized through the Brundtland Commission with its arguments for sustainable development through attention to the links between protection of natural systems and strategies for economic development.

At the strategic level, there is a growing effort to develop alternative ways in which integration can be realized. However, at the operational level, where interest turns to what will be done, there have been mixed results. Practical matters regarding how many aspects should be integrated, how to overcome the ever present edge or boundary problems and how to overcome professional biases and suspicions still often represent major barriers against effective integrated water management.

2. The scope of integrated water management

Initial attempts to interrelate water with other considerations seem to have had several characteristics. First, the early initiatives usually concentrated upon only a few concerns, such as soil erosion, flood control and drainage, or hydroelectricity generation, irrigation and flood control. Second, subsequent attempts were made to broaden the range of concerns to be addressed. In that respect, many management functions were identified and then allocated to become the responsibility of an implementing agency. However, almost inevitably, a much narrower range of functions tended to be pursued, usually as a result of financial arrangements which favoured some matters over others, or because of professional pre-dispositions to concentrate upon selected activities. Third, a stage would be reached when it was realized that it was extremely difficult to implement a wide range of management functions in a comprehensive manner. The response then was often to focus on a narrower range of considerations, although to consider explicitly the linkages among them. Whichever approach was adopted, reliance was placed upon organizations in the public sector to expedite management of water.

Experience with the seven countries illustrates this pattern. In the United States, early efforts at integrated management led to consideration of navigation, flood control and drainage in the eastern states and to hydropower generation, irrigation and flood control in the western states. As pollution and water quality questions emerged onto the public agenda in the 1970s, efforts were made to incorporate quality with quantity issues. The most common strategy became to consider multiple purposes through public agencies with reliance upon structural means. The emphasis upon structural or technological measures which focus upon manipulation of the natural system tended to preclude consideration of non-structural approaches which emphasized modification of human behaviour. During the 1980s, the trend in the United States moved steadily away from an integrated approach as various co-ordinating mechanisms were eliminated. The outcome has been a general strategy to tackle problems in a more single-minded manner rather than considering the linkages with other issues.

This pattern appears in other countries as well. For example, in Japan the primary attention has been upon irrigation to support food production, hydropower to help a nation lacking its own oil and natural gas reserves, and flood damage protection to reduce potential harm to the high population density settlements. It was not until after the Second World War that attention was turned to the growing pollution problems associated with an increasingly industrialized and urbanized society. The Japanese realized fairly early that it was not necessary to rely exclusively upon the public sector to address these problems. The example of the Yahagi River Basin Water Quality Protection Association shows how interest groups from the private sector can combine their interests and talent to tackle problems that in many countries are left to the public sector.

With the establishment of eleven River Basin Development Authorities in Nigeria, responsibility for a broad range of management functions was allocated to the new authorities. These functions cover urban and rural water supply, control of floods and erosion, irrigation and drainage, fisheries and navigation, forestation and afforestation, pollution control and resettlement. Given the overriding importance of growing food for a rapidly growing population and of providing potable water to that population, it is not surprising that attention has concentrated primarily upon irrigation and water supply projects. Relatively little attention has been given to soil conservation, flood damage reduction and forestation. Thus, while there is authority for wide-ranging management activity, a conscious decision apparently has been taken to focus upon a narrower range of activities.

In England, the early concern was with water supply and water quality. Over time, there has been an evolution towards a smaller number of river basin authorities which have received a steadily larger number of management functions. As we begin the 1990s, the Thatcher government is going through the process of 'privatizing' some of the functions (water supply, sewage treatment) provided by what have been public authorities. Some have described the move to privatization as a breaking down of a system which has sought to realize integration by centralizing most of the key water management functions in ten river basin authorities. Others have observed that the division of functions between the public and the private sector does not have to represent a breaking down of an integrated approach—but rather, represents a further move toward integration by combining the talents of the public and the private sector.

In most of the countries examined in this book, individual agencies and organizations have been expected to develop and shape their own goals and directions. Once that has been accomplished, there has been a search for the most appropriate co-ordination mechanism to ensure that the individual goals and directions are integrated to improve societal welfare. Only in a few instances, such as Poland, where there is more centralized planning, has there been a specific attempt to identify more general development goals and then to construct strategies for water management which will support the more general goals. Whichever approach has been taken, the task of establishing overall societal goals is often frustrated by the existence of different but legitimate interests in societies at any given time. This situation stresses the importance of developing means to facilitate the comparison and balancing of different hopes and aspirations within a society.

3. Boundary problems

Boundary or edge problems arise in situations in which two or more interests overlap. For example, this may occur when one agency of government wishes to preserve or expand wetlands in order to enhance wildlife habitat whereas another agency wishes to drain them in order to increase agricultural production. It may arise when one agency proposes building a dam to facilitate the generation of hydroelectricity while another agency expresses concern that the dam may damage migratory fish. If these interests are to be 'integrated' into a systematic management strategy, it is essential to develop mechanisms through which different values and interests can be identified and articulated, and then decisions made which reflect these different and often conflicting aspirations.

Each of the seven countries examined here has to deal with boundary problems. At the federal level in the United States, key agencies are the US Army Corps of Engineers, the Bureau of Reclamation and the Soil Conservation Service. Overlap also emerges between the federal and state levels of government. New Zealand does not have states or provinces, but boundary issues emerge among the line agencies of central government, catchment boards and regional planning authorities. In Canada, jurisdiction over water is shared between the federal and provincial governments, ensuring that much water management must be conducted as a joint effort between those two levels of government.

In England, the ten regional water authorities have had to balance their goals and activities with those of line agencies of the central government as well as with the municipalities. In Japan, the Special Area Multiple Purpose Development Projects created in 1950 were used to try and bridge the interests of central government agencies and the prefectures. The Kitakami River Multiple Purpose Development illustrated how this approach has been used to combine a number of water management functions and the interests of different participants.

The central government has been a strong force in water management in Poland. However, with the creation of provincial governments the need has arisen to address the respective responsibilities of the two levels of government. The experience in Nigeria shows how difficult it can be to integrate the concerns of a central government with those of the states. In that country there have been major problems, especially when the elected political party in the state has been a different party from the one which leads the central government.

It is apparent that governments in these different countries have recognized that boundary problems cannot be overcome simply by rearranging structures. There certainly has been reorganization of government structures over time, but that has not been the only device relied upon to achieve co-ordination and integration. In the following section, attention turns to some of the methods which have been used to achieve integration.

4. Instruments for integration

Before highlighting the instruments used to achieve integration, it is important to emphasize the *context* within which water management is conducted in each country. Awareness of the context will help in making any judgements regarding the feasibility of transferring experience from one country to another.

The United States is a large country with considerable hydrological diversity. The country is often described as humid in the East and arid in much of the West. It has a federal system of government. Muckleston suggested that ideology has been a key factor in the development of water management strategies, and indicates how strategies have shifted in accordance with changes in prevailing ideology.

New Zealand provides a sharp contrast to the United States. It is a relatively small nation in both area and population. The catchments are relatively small, with less than 50 per cent of the river basins being over 1,500 square kilometres. Over half of the population is located in urban areas, with about one-third of the population living in the capital city of

Auckland. The current government is Labour. Since 1984 there has been a strong free enterprise policy which has been the driving force behind decisions being taken to restructure the way in which resource management is handled.

Canada has an abundance of water in a large country with a relatively small population. However, much of the supply is oriented toward the North of the country while the majority of the people live in the South. There are problems of water shortages in selected regions in the South, as well as flooding problems in some communities. On the quality side, toxics and nonpoint source pollutants represent serious problems. Under the Canadian constitution, the federal government and the ten provincial governments share responsibility for many aspects of water management and development.

England is a nation small in area with a relatively high population density. With the Industrial Revolution beginning in that country, it now has a heritage of outdated and failing supply and waste treatment infrastructure. In addition to having to deal with the capital costs of refurbishing the basic water infrastructure, the water agencies are concerned about regulation of effluent discharge and nonpoint sources of pollution. Between the 1950s and 1980s, there was a shift away from the welfare state approach towards a market-based approach. It is in this country that major initiatives are being taken to privatize selected aspects of public water agencies.

Japan must deal with extreme spatial and temporal variations in rainfall, and during the typhoon season must deal with the possibility of extreme damage from storms reaching the country from the Pacific Ocean. The population density is very high, and the associated high degree of industrialization places considerable stress upon aquatic resources. The central government manages water through a substantial number of complex statutes, but also often utilizes joint approaches with the prefectures or county levels of government.

For Poland, surface water is the major source of water, but it is spatially and temporally irregular. Following the Second World War, attention was directed to developing water supply systems so that domestic and manufacturing needs could be met. For a long period, water quality problems were not addressed, and these have now emerged as a major issue. Poland experienced a serious recession in the 1970s, and currently is entering a new phase in the type and style of government being used. The evolution of style of government in Poland creates considerable uncertainty, and it is in this context that strategies are being developed for water management. It seems clear that water management objectives will be set with regard to plans for economic and industrial expansion.

There are some parallels between Nigeria and Poland, with each country having a majority of the population dependent upon agriculture for its livelihood, and the standard of living for many people still being relatively low. Like Japan, Nigeria experiences significant variation in rainfall. Shortages of water are common in the north of the country, where irrigation projects receive priority. Improvements in domestic water supply systems for both urban and rural dwellers is a priority. Water management in general is conducted with a lack of reliable information for much of the country regarding both quantity and quality. Financial and technical resources also provide a constraint on the range of strategies which can be considered.

It is against this diversity of contexts that the following comments about instruments to achieve integrated water management should be considered.

4.1. Legitimation

The countries examined here have relied upon a mix of legislation, political commitment and administrative decisions to foster integrated water management. Not all of these instruments have been focused directly upon water, but their effect has often been significant for integration.

In the United States, the Water Resources Planning Act in 1965 was designed to improve co-ordination among federal agencies and to improve project efficiency. At the same time, it made provision for more planning activity to be conducted by the states, thus fragmenting the effort in water planning. The National Environmental Policy Act in 1969 served to require federal agencies to consider explicitly the environmental consequences of proposed activities. Regarding water management, NEPA contributed to pushing federal agencies to incorporate environmental considerations along with those of a technological and economic nature into their decisions. At a more restricted level, significant progress has been made under the unified national programme for floodplain management in which both structural and non-structural adjustments are integrated to reduce flood damages. This programme draws upon a number of statutes and administrative decisions for its legitimacy.

New Zealand passed the Soil Conservation and Rivers Control Act in 1941. That statute provided the basis for comprehensive catchment planning and management with regard to soil erosion, flood control and land drainage. This statute and the Water and Soil Conservation Act in 1967 provided the basis for programmes at the national level. At the local level, the two acts pertinent for water and land management were the Local Government Act in 1974 and the Town and Country Planning Act in 1977. As with NEPA in the United States, the two locally-oriented statutes in New Zealand indicate how legislation conceived for a variety of purposes may combine to give legitimacy to an integrated approach.

The division of responsibility for water between the federal and the provincial governments in Canada provokes uncertainty as to who is responsible for what, creating an obstacle for an integrated approach. However, the Canada Water Act in 1970 helped to define the way in which the federal and provincial governments could combine their efforts to address water quality problems. It was not until 1987, however, that the federal government developed a systematic and broad-ranging water policy. That initiative should help to clarify roles and thereby contribute to more effective integration. Individual provinces have used legislation to foster an integrated approach. Probably the best known is the Conservation Authorities Act of 1946 in Ontario. While that provided authority for one agency to be responsible for many aspects of renewable resources, government restructuring in 1972 split water quantity and quality aspects between two provincial line ministries in Ontario. That action appears in other countries as well—when a government takes one action to enhance an integrated approach and then another action which undercuts that approach.

In England, a series of acts have been used to establish catchment-based authorities and to increase gradually their range of responsibilities. The British tradition has been to rely more upon discussion and persuasion than upon legalistic measures, however, so many actions do not have a statutory basis in that country. The Water Act of 1983 certainly had a business or market-oriented approach, and that was followed by a Water Bill in 1989 which is the basis for privatization. The initiative for privatization is strongly backed by the ideological commitment of the ruling Conservative Party, so in this instance both political and legal supports are being used.

In Japan, a variety of legislation has been used for water management and development, to the extent that the number and complexity of the acts have served as an obstacle to integration. Perhaps in recognition of the inertia created through the host of legislation, the government moved to develop a national comprehensive water resources plan in 1987. Thus, through an administrative measure of a planning exercise it was hoped to ensure a greater measure of integration than was possible to realize through the legislative route.

Poland introduced a water law in 1979 to cover a wide range of water management functions: supply, navigation, hydropower generation, pollution control. Prior to the legislation, it also developed a perspective plan for water management during 1975. With the plan and the legislation, it is hoped that the various dimensions of water management will be drawn together into a coherent package.

For Nigeria the major initiative was the creation of the eleven River Basin Development Authorities in 1979. The government decree under which the authorities were established provides them with many responsibilities for planning, management and development. However, as already noted, due to the presence of different political parties in the central government and some state governments, there has been resistance experienced in implementing some of the programmes.

The seven countries under review here have relied heavily upon legislation to provide legitimation for integrated approaches. It is apparent that when the statutes are combined with political commitment, steady progress is realized either toward or away from an integrated approach. The ideology of a President Reagan, it has been suggested, has driven the approach to water management in the United States away from integration. In England under the forceful leadership of Prime Minister Thatcher, a strong push is being made toward privatization. At the same time, legislation by itself does not seem sufficient to achieve integration. Every country considered in this book has one or more acts which could facilitate integration, and yet each one has also encountered difficulties in achieving effective integration.

4.2. Functions

In the analysis of experiences in the seven countries, attention was focused primarily upon substantive rather than generic functions. The general pattern is for a small number of functions to be considered together, while the large number of remaining functions appear to receive secondary attention. As will be noted in Subsection 4.3, the various functions are usually distributed among a number of line agencies for implementation. Only in a few instances have a large number of functions been delegated to a single agency.

Navigation, flood control, hydroelectric power generation and irrigation seem to be the primary functions in the United States which have been considered jointly. Questions associated with domestic and industrial water supply, waste treatment, leisure and wildlife have been less prominent when efforts have been made to integrate a variety of substantive functions.

In New Zealand, initial emphasis was upon soil erosion, flood control and land drainage. However, under the Water and Soil Conservation Act of 1967, provision was made to consider the interrelationships among water quality, soil conservation, flood protection, water supply, power generation, recreation, wildlife and scenic values. Not all of these have been given equal attention in practice, and the ongoing changes in government organization makes it difficult to foresee which aspects will be emphasized in the future.

It is difficult to generalize about the situation in Canada, since the federal and provincial governments take different initiatives at different periods. However, it is probably fair to say that there has been a movement from reliance upon single purpose and single means strategies to multiple purpose and multiple means strategies. The allocation of different functions to the federal and provincial governments can lead to overlap and duplication of effort. Given these qualifications, there have been instances where water supply, pollution, flood control and water-based recreation have been considered together. Hydropower generation and irrigation tend to be implemented as single purpose, single means projects in Canada. As in the United States, there has been increasing consideration of structural and non-structural adjustments in floodplain management, and currently demand management methods are receiving attention.

Water supply and waste treatment have been the dominant functions in England, with conservation, fisheries, navigation, flood control, drainage and recreation receiving secondary attention. While water supply and waste treatment have been of primary concern, there has been a long tradition of separating the responsibilities for clean and dirty water in England among different organizations. One benefit of privatization may be that as water supply and waste treatment are the two functions to be 'sold' to the private sector, it may lead to their being handled in a more integrated manner than has been the case in the past.

Water management has been primarily geared to agriculture, industry and urban needs in Japan, with hydropower generation also receiving significant attention. Sewage treatment has often been inadequately dealt with, and currently water quality is a major concern. There has been a tendency to rely upon a structural approach, with dams being a dominant element in management and development strategies. In instances such as the Kitakami River we can see that several functions such as flood control, irrigation and power generation have been handled in an integrated manner.

In Poland, legislation passed in 1974 provides for water management at the central government level to consider water supply, flood damage prevention, hydropower generation and water rights. As in other countries, there is still difficulty in ensuring the joint consideration of surface and groundwater resources. At a generic level, planning and development are led by the central government whereas licensing is the responsibility of the 49 provinces.

Primary emphasis in Nigeria has been given to domestic and industrial water supply and to irrigation projects. These functions have been given to the eleven River Basin Development Authorities. These authorities have also received other substantive functions, such as flood and erosion control, fisheries and navigation, power generation and forestation. However, there has been a preoccupation with supply and irrigation aspects and relatively little attention given to the other functions.

The experience in the seven countries indicates that a comprehensive approach, where that means a systematic consideration of a wide range of functions, has rarely been implemented. Instead, the approach almost invariably has been to concentrate upon a smaller number of functions, the specific ones reflecting conditions and concerns in each specific country. And, the mix of functions to be combined may vary from one region to another within a country, particularly if the country is large enough to have significantly different physical, settlement and economic conditions from region to region.

4.3. Structures

A number of observations can be made regarding structural arrangements. Many countries have created an organization to co-ordinate water management at a national scale. In the United States, the National Resources Planning Board was used to address land and water issues, and the Water Resources Council was used to examine various aspects of water. In New Zealand, a National Water and Soil Conservation Authority was created and in England a co-ordinating organization was formed from the chairmen of the regional water authorities. In contrast, countries such as Canada have developed no effective mechanism to integrate the different dimensions of water at a *national* scale.

At a regional level, the structural arrangements have generally taken one of two forms, but occasionally have been a combination of the two forms. In a relatively few instances, river basin authorities have been established and given responsibility for a mix of functions. Perhaps the best known is the Tennessee Valley Authority, yet it has not been replicated anywhere else in the United States. One of the reasons for lack of replication is alleged to be the determination by other line agencies to stop the introduction of any more catchment authorities which would be given powers normally held by the line agencies. The Americans have also used interstate commissions such as the Delaware River Basin Commission and the Susquehanna River Basin Commission along with more general River Basin Commissions. In New Zealand, catchment boards or water boards have been introduced to handle problems at the regional level, but, as in the United States, they have never covered the entire country.

In Canada, watershed-based agencies have been established in Ontario and Manitoba, while in Alberta the province has been divided into six catchment areas for the purpose of planning. Furthermore, Canada and the United States have jointly worked through the International Joint Commission to deal with problems of lake level fluctuations and water quality in the Great Lakes.

England has developed the most sophisticated set of regional water authorities. Ten such organizations cover England and Wales, and they have been given a wide range of management functions. In contrast, in Japan water management is dominated by the central government and municipalities. Only in selected areas such as the Kitakami River have organizations modelled after the Tennessee Valley Authority been put in place. Poland relies heavily on the central government line agencies whereas in Nigeria authority has been delegated to the eleven River Basin Development Authorities.

Several conclusions can be drawn for the regional experiences. The catchment area is often used as the unit for management. In many instances, existing line agencies are expected to implement their programmes in the context of river basins. It is hoped that through co-ordinating mechanisms these different line agencies will be able to integrate their activities. Alternatively, catchment-based authorities have been created, as in England, New Zealand and Nigeria.

Where catchment-based authorities have been put in place, the practice appears to be that only a small number of management functions are handled, even though a broader range of functions has been delegated to them. This happens usually because the catchment-based organizations have not been given sufficient authority to carry their programmes through or as a result of resistance from existing line agencies. Certainly there does not seem to be a move toward setting up 'super agencies' based on catchment units.

In general, there appears to be a trend toward decentralization rather than centralization, with reliance upon a greater number of specialized agencies rather than the use of centralized organizations with many functions. In some ways, concern about the most appropriate form of structures does not address the fundamental issue of boundary or edge problems. Overlap of interests and responsibility always exists. The question becomes whether such overlap should be internalized within a larger organization, or externalized among a larger number of small organizations. No matter which option is selected, the boundary problems do not disappear. Rather, they appear in different versions.

4.4. Processes and mechanisms

It was indicated in the first chapter that there is never a perfect 'fit' among legitimation instruments, functions and structures. As a result, use is made of various processes and mechanisms to overcome the problems which occur because of imperfect matches. It is often these processes and mechanisms, informal and formal, which facilitate co-ordination and integration.

The conclusion from the seven countries is that there is a lack of effective processes and mechanisms to facilitate integration. The lack of such processes and mechanisms was identified as explicit weaknesses for the United States, Japan, Poland and Nigeria, and problems also seem to be present in New Zealand, Canada and England. It appears that development of better processes and mechanisms to expedite integration may be one of the greatest needs at the moment.

A variety of inter-departmental or inter-agency committees has been used. In the United States, there have been Federal Interagency Committees, and the Northwest Power Planning Council. In New Zealand, there have been various committees to deal with regional planning, catchments, and land use. Canada has used an inter-departmental water committee at the federal level, and has a number of federal–provincial committees to implement programmes ranging from water quality to flood damage reduction. Different committees are used in England to coordinate the activities of central government line agencies, and to coordinate the programmes within regional water authorities.

A variety of specific mechanisms or tools is used to encourage integration. In the United States, the National Environmental Policy Act created a process which stipulates that environmental considerations must be incorporated into management decisions. In countries such as Canada, environmental concerns are handled by an administrative mechanism at the federal level rather than by legislation. Development plans in Japan, Poland and Nigeria serve as a means to force consideration of water management issues relative to other economic, social and environmental concerns.

Some very specific tools have also been used. In New Zealand, by-laws, water rights, water allocation plans, conservation orders and water management charges comprise a mix of regulatory mechanisms which force consideration of a range of issues. Non-statutory mechanisms in that country include water quality surveys, as well as soil and water management plans. Cost-sharing arrangements, accords and regulations are used in Canada. It appears as if the Canadians will also develop an explicit pricing policy which may help to integrate some aspects of water quantity and water quality.

Abstraction licences, discharge consents and river quality objectives are used in England. In Japan, the process of designating rivers into qualitative categories based on economic and social significance forces a variety of concerns to be addressed. Water supply and waste management charges are used in Poland to influence water uses and their impacts.

Different strategies to combine a top down with a bottom up approach are also being used. In the United States and Canada, a variety of public participation procedures has been used to involve the public in management decisions. In England, there has been a shift away from involving the public in actual decisions as the regional water authorities have been streamlined to make them more 'businesslike'. At the other extreme are organizations such as the Yahagi River Basin Water Quality Protection Association in Japan. The YWPA was formed in 1969 by eighteen groups and its tasks are to maintain water quality relative to agreed upon standards and to assess the appropriateness of proposed large-scale development. It appears as if much could be learned from careful examination of the experience with interest-based management organizations in Japan. Such organizations provide a useful counterbalance to overreliance upon public agencies.

4.5. Organizational culture and participant attitudes

Explicitly identifying the organizational culture and the participants' attitudes in a country is difficult, since that which is stated and that which is done may not always be identical. In other words, policy documents and guidelines in most countries will urge co-ordination and co-operation, and respondents' public position usually will support the concept of integration. However, in practice, individual agency representatives will often attempt to obtain an advantage for their organization, and individuals will often try to enhance their performance image even if that is at the expense of the collective welfare. The attitude will vary from culture to culture, but generally the tendency is not toward integration and co-operation.

Experiences in a number of countries covered in this book verify this conclusion. In the United States, it has been suggested that key positions have been given to individuals who share a common ideology which minimizes a concern for integration. In New Zealand, it has been argued that there were often fundamental differences between water engineers and soil conservators, leading to professional rivalry and outright conflict. More recently, there has been sharp conflict between older managers on the Water Boards and more newly trained managers who have stressed business and profit considerations. Differences have been observed in England between those who have stressed technical efficiency and those who have emphasized economic efficiency. It has also been suggested that some central line agencies have been designed almost to ensure that they will be remote and inaccessible, attributes unlikely to foster integration and co-operation with other agencies. Nigeria is characterized as having a complex bureaucracy, as well as political parties which are often as concerned with scoring political points against different parties at other levels of government as they are with providing useful processes and outputs.

On balance, therefore, there is evidence to suggest that many institutional arrangements seem to reinforce a tendency away from integration and more towards enhancing self-interest. On the other hand, some experiences suggest that there are also attempts at developing and cultivating a spirit of co-operation and integration. Using a bottom up approach, in the Yahagi River basin in Japan connections have been made between upstream and downstream communities through exchanges of schoolchildren and information about economic activities of the respective towns. Using such exchanges as well as general education programmes, the goal has been to create a sensitivity about living in a common environment, and to cultivate an attitude about responsibility for other people living and working in the same catchment.

At a bureaucratic level, increasing attention is being given in countries such as the United States and Canada to develop mechanisms for mediating, bargaining and negotiating. This approach has long been used in

England. The goal is to move away from adversarial confrontations in which there are declared 'winners' and 'losers' towards situations in which many interests can conclude that they gained through joint efforts to resolve problems. Not every situation lends itself to all people concluding that they have won, but increasingly the belief seems to be that there is enough to gain that it is worthwhile cultivating negotiating skills.

As noted in the first chapter, ultimately it is people who make a system work, or cause it to falter or stumble. While the documentation of organizational culture and participants' attitudes is still incomplete and weak, it does appear that these aspects should receive significant attention in the future. It may well be that improvements in these aspects as well as in processes and mechanisms will offer the greatest likelihood for significant advances regarding integration.

5. So what?

Experiences from a selected number of countries must be used cautiously in drawing out general implications. Values may differ from culture to culture, and physical and economic conditions vary from country to country, as well as within countries. Notwithstanding this obvious qualification, it appears that there is substantial agreement that it is desirable to use an integrated approach in water management, where that may mean linking different components of water, relating water to land and environmental resources, or incorporating water into social and economic development.

At a conceptual level, integration receives considerable support. At an operational or working level, there are numerous problems in implementing the concept. A major issue is the existence of boundary or edge problems that are always present, even though their nature may change over time.

The suggestion from the countries examined here is that there is no one tool or instrument which will expedite integration. Legitimation is often sought through legislation, but there are many examples of statutes not being used or of one statute nullifying the impact of another. For integration to occur, a high level of political and bureaucratic commitment is needed in addition to legislation.

Functions may be generic or substantive. The pattern to date appears to be the integration of a small number of substantive management functions, either through one agency or through a number of agencies. There is a positive aspect to this form of integration, in that goals and objectives can usually be kept in sight and performance targets can be pursued. At the same time, it is desirable not to lose sight of the broader range of functions which need to be considered. In this respect, it will be appropriate to pursue a strategy in which a continuous scanning of a wide range of functions is maintained even if a smaller number is being concentrated upon for implementation.

Too often, it appears as if reorganization of structures is used to try and achieve integration. It is clear that reorganization never eliminates the boundary problems, but simply redefines where the boundaries will be. In only a few situations have larger, more centralized, multiple function agencies been used. More frequently, the pattern seems to be to rely on line agencies to implement those aspects of an integrated strategy which fall within their mandates. This approach requires that attention be given to processes and mechanisms which can get around the boundary problems.

While it appears as if processes and mechanisms are crucial for successful integration strategies, experience indicates that no single instrument by itself is adequate. Indeed, while often the conclusion is that there is a need for greater co-ordination and co-operation, there is also a great need for more systematic evaluative research to assess the strengths and weaknesses of various instruments in the context of different situations.

Since people are the ones who make an integrated approach succeed or fail, much more attention needs to be directed to cultivating an organizational culture and attitudes that reinforce the benefits to be derived from integration. At the moment, most of the incentives encourage less rather than more integration. Development of skills in bargaining, negotiating and mediation will have to be central to improving the atmosphere for more effective integration.

Integration is not a panacea for water management problems. Indeed, it is not an end in itself, but rather it is a means to the end of more effective resource management. While approaches must be designed to fit the conditions specific to a country, it appears that ideas can be developed from examining experiences in other jurisdictions. The intent of this book has been to present a diversity of experiences and a framework to facilitate their comparison. It is left to the reader to determine which aspects of these experiences may be helpful in resolving problems that he or she must address.

Index

accountability 12, 78, 96, 98
agriculture 1, 11, 29, 36, 52, 89, 91, 120,
    139, 143, 149, 154, 157, 164, 165, 173,
    188, 190, 193, 194, 206, 208, 212
Aichi Prefecture (Japan) 164
Aichi Yosui Project (Japan) 154
Alberta 89, 102, 105, 108, 213
Aluminum Company of Canada 108
aluminum smelter 108
anglers 127
antagonism 15
arbitration 8
Arctic 91
Arkansas—White Basin (United States) 34
Auckland 46, 58, 72, 76, 208
Australia 3

baiu 149
Bakalori project (Nigeria) 198
balanced use 67
Balkanized approach 41
Baltic Sea 172, 185
bank protection 31
bargaining 8, 13, 15, 113, 139, 216
Basic Law for Environmental Pollution
    Control, 1967 (Japan) 158
Bay of Plenty 68
beneficial use 67
benefit cost analysis 14, 56
Benin–Owena River Basin Development
    Authority (Nigeria) 200
Berger Commission Inquiry (Canada) 109
biological oxygen demand 166
biosphere reserves 185
Bonneville Power Administration (United
    States) 37
bottom up 14, 17, 215, 216
bounded holistic perspective 16
Britain 194
British Columbia 89, 102, 105, 108
British Waterways Boards 133
Brundtland Commission (World
    Commission on Environment and
    Development) 1, 111, 170, 204
Bureau of Reclamation (United States) 25
by-laws 54, 57, 64, 129, 215

Canada 88, 203, 206, 208, 209, 211, 213,
    214, 215, 216
Canada Water Act, 1970 99, 102, 112, 210
Canadian Council of Resource and
    Environment Ministers 111
Canadian Environmental Protection Act
    103
Canadian Heritage River System 102
Canadian Wildlife Federation 107
Canterbury 68
Carter administration 40
catchment boards 49, 54, 60, 63
catchment management 24, 30, 33–37, 50–
    57, 148, 158, 187, 214
Central Arizona Project (United States) 28
centralization 12, 51, 78, 214
charges 64, 215
climate 193
climate change 113
Clutha River 67
coal 48
coastal areas 39, 91, 98, 105, 135, 158, 173
codes for buildings 40
coefficient of river regime 150
collaboration 97
Colorado Basin (United States) 31
Columbia Basin (United States) 34, 36–37
communication 168
compensation 152, 164, 198
Comprehensive National Land Development
    Act, 1950 (Japan) 153
compromise 8, 15, 16, 78, 92, 93, 139
Confederation of British Industry 132
conflict resolution 97, 112
conflicts 8, 163
confrontation 93
consensus 35
consent system 129
conservation 24, 60, 133, 177, 195
Conservation Authorities, Ontario 3, 93–
    98, 112, 203
Conservation Authorities Act, 1946
    (Ontario) 94, 210
Conservation Districts (Manitoba) 98
Conservative Party (Britain) 121, 127, 210
Conservation Reports (Ontario) 96, 112

Conservation Strategies 110
consultation 139
context 8, 207
Control of Pollution Act, 1974 (England)
    131, 141
co-operation 14, 62, 112, 154, 216
co-ordination 4, 6, 10, 14, 28, 29, 30, 33, 34,
    39, 40, 59, 136, 139, 199, 206, 216
co-ordination trap 9
Corps of Engineers (United States) 25, 36,
    37
Council for the Preservation of Rural
    England 134
Council on Environmental Pollution
    Prevention (Japan) 168
counterenvironmentalism 28, 29
Cracow 176
crisis-oriented 55
cult of co-ordination 7

dams 31, 153, 196–197, 198
decentralization 12, 62, 214
Delaware River Basin Commission (United
    States) 35, 213
demand management 125
Democrats (United States) 24, 26, 27
demonstration work 52
devolution 28, 29, 78, 84
dilution 129, 175
dirty water 134–135
Disaster Protection Act, 1973 (United
    States) 41
disease 192
disinformation 15
double cropping 193
drainage 1, 11, 30, 50, 52, 53, 55, 59, 63, 68,
    76, 82, 92, 97, 98, 129, 133, 139, 157,
    190, 191, 193, 195, 204, 205, 206, 209,
    211, 212
drought 152, 155, 169

economic development 1, 119, 178, 184
economic growth 127, 192
economies of scale 12, 30, 96, 98, 135
ecosystems 1, 110, 168, 184
effluent 66, 76, 128, 129, 157
elitist political culture 144
energy 36, 105
enforcement 39, 133
England 2, 119, 148, 203, 205, 207, 208,
    210, 211, 212, 213, 214, 215, 217
environmental deterioration 26
environmental impact assessment 14, 28,
    104–109, 158, 166
environmental movement 26
environmental quality 27, 29
erosion 11, 46, 50, 51, 52, 53, 57, 59, 60, 67,
    72, 76, 78, 94, 97, 191, 195, 203, 204,
    205, 209, 211
European Economic Commission 48

eutrophication 158
evapotranspiration 191
externalities 139

farm conservation plans 57
farmers 127, 141
Federal Emergency Management Agency
    (United States) 39
Federal Energy Regulatory Commission
    (United States) 32
Federal Environmental Assessment and
    Review Office (Canada) 105
Federal Environmental Assessment Review
    Process (Canada) 105–109
Federal Power Commission (United States)
    32
federalism 91, 112, 113
fertilizer 36, 120
financial 8, 96, 98, 112, 200, 204
fire 50
First National Development Plan, 1962–1968
    (Nigeria) 193, 194
fiscal restraint 113
fisheries 1, 31, 32, 33, 36, 46, 58, 60, 92, 98,
    105, 120, 129, 133, 157, 165, 184, 195,
    198, 200, 205, 212
flexibility 13
flood control 1, 11, 30, 31, 33, 34, 36,
    38–41, 50, 73–75, 139, 160, 204, 209,
    211, 212
Flood Control Act, 1936 (United States) 38
Flood Disaster Protection Act, 1973 (United
    States) 39
flood warning system 94
flooding 46, 51, 53, 55, 59, 60, 63, 68, 70,
    72, 73, 74, 89, 94, 97, 107, 149, 152, 156,
    160, 163, 166, 169, 173, 182, 190, 191,
    195, 203, 205, 208, 215
Food and Agricultural Organization 194
food production 153
forestry 11, 36, 46, 52, 78, 80, 91, 176, 195,
    198, 205, 212
fragmentation 6, 10, 14, 38, 40, 57, 91, 92,
    161, 169, 176, 192, 203
France 3
Fraser River 102
Fredericton 89
Free Trade Agreement 110
functions 10–11, 211–213
Fund for Water Management (Poland) 178

gas 48
Gongola Basin (Nigeria) 198, 199
goodwill 14
Grand River Conservation Authority
    (Canada) 98
grassland farming 46
Great Lakes 91, 108, 111, 213
green belt 135
green movement 186

groundwater 1, 40, 64, 91, 124, 148, 152,
    156, 169, 173, 175, 177, 184, 185, 191,
    195, 203, 212

Hamilton 72
harbour degradation 72
health 58, 103, 119, 139, 157
Hokkaido 149
holistic approach 16, 38, 45, 119, 133, 136
Honshu 149
Hoover Dam (United States) 31
horticulture 46, 164
hydroelectricity 1, 8, 11, 25, 26, 31, 32, 36,
    48, 67, 92, 97, 108, 149, 153, 154, 160,
    161, 163, 169, 176, 195, 203, 204, 205,
    206, 210, 211, 212
hydrological cycle 169
hydrology 178

Ichinoseki City 160
Idaho 32, 36
ideology 8, 13, 23–30, 78, 142, 144, 207,
    211, 216
India 3
Indian Reserves 92
individualism 24
Industrial Revolution 208
Industrial Water Law, 1950 (Japan) 156
Industrial Water Works Law (Japan) 156
industrialization 36, 127, 154, 179, 184, 185,
    188, 205, 208, 212
infiltration 191
informal mechanisms 13
inland waterways 31, 36
institutional arrangements: defined 6–7
insurance 73
integrated water management: scope 4
inter-agency rivalries 109
interbasin transfers 91, 108, 110, 120, 121,
    124, 137, 148
International Joint Commission 213
interstate compacts 34
intervention 8
inviolable runoff 173
irrigation 24, 26, 31, 53, 68, 92, 152, 157,
    160, 161, 163, 164, 176, 179, 184, 185,
    191, 193, 194, 195, 198, 200, 204, 205,
    209, 211, 212
Ishikari River 150
Iwate Prefecture (Japan) 158

James Bay 108
Japan 148, 203, 205, 207, 208, 209, 210, 212,
    213, 214, 215

Kaduna 192
Kano River project (Nigeria) 198
Kemano Diversion (Canada) 108
Ken 154
Kiri Dam (Nigeria) 198, 199

Kiso River 154
Kiso special area 153
Kitakami River 149, 213
Kitakami River Multiple Purpose
    Development (Japan) 158–164, 207
Kitakami special area (Japan) 153
Kyushu 149

Labour Government (New Zealand) 45, 78,
    84, 208
Lagos 192, 199
Lake Erie 91
Lake Ontario 91
Lake St. Clair 91
land acquisition 39
Land and Water Management Plan (New
    Zealand) 76
land capability 55
land degradation 203
land development 153
land improvement districts 157
land subsistence 72, 156
Land Use Capability Survey (New Zealand)
    56
land use management 39, 40, 54, 57
landowners 15, 127
landslide 73, 149
Law Concerning Special Measures for
    Conservation of the Environment of the
    Seto Island Sea, 1973 (Japan) 158
legal 8, 93, 148
legislation 9, 211
legitimation 8–10, 209–211
licensing 35, 156, 177, 182, 215
litigation 93
livestock 120
local communities 53, 94, 154, 156, 184, 186
local government 126
Local Government Act, 1974 (New
    Zealand) 50, 71, 76, 209
localized approach 123
London 135, 138, 142
low flow augmentation 97

MacKenzie Valley Pipeline (Canada) 109
Manawatu region (New Zealand) 52
Manitoba 89, 93, 98, 102, 105, 107, 108, 213
manufacturing 46
Maori 46, 62
market principles 11, 23, 28, 29, 84, 119,
    139, 140, 142, 208
mediation 13, 216
Meijiyosui Land Improvement District
    (Japan) 164
Miami Conservancy District (United
    States) 2
minerals 91
minimum acceptable flow 129
Mississippi 30

Mississippi River Commission (United States) 31
Missouri Basin (United States) 26, 31, 34
Miyagi Prefecture 158
monitoring 108, 168
Montana 36
Montreal 102
Motu River 68
Municipal Industrial Strategy for Abatement (Ontario) 103
Muskingum Watershed Conservancy District (United States) 2
myth of super abundance 89

Nagoya 152
National Comprehensive Development Plan, 1962 (Japan) 155
National Comprehensive Water Resources Plan (Japan) 148, 210
National Conservation Strategies 110
national economic development 27, 29
National Environmental Policy Act, 1969 (United States) 27, 209, 215
National Flood Insurance Program (United States) 39
national forests 24
National Land Agency (Japan) 148
national parks 92, 135
National Programme for Natural Environment Protection to the Year 2010 (Poland) 182
National Resources Planning Board (United States) 26
National Rivers Authority (England) 121, 128, 134, 143
National Task Force on Environment and Economy (Canada) 111
National Water and Soil Conservation Authority (New Zealand) 60, 74
National Water and Soil Conservation Organization (New Zealand) 60
National Water Commission (United States) 27
National Water Council (England) 140
nationalization 127
natural hazards 73
natural values 40
Nature Conservation Council (New Zealand) 58, 133
navigation 11, 30, 31, 32, 36, 92, 133, 176, 195, 198, 200, 205, 210, 211
Nechako River 108
negotiation 5, 13, 15, 93, 104, 216
New Brunswick 91, 101, 102, 105
New Deal 25, 36
New River Law, 1964 (Japan) 155
New York – New England Basin 34
New Zealand 3, 45, 203, 206, 207, 209, 211, 213, 214, 215, 216
New Zealand Catchment Association 64

New Zealand Institute of Engineers 59
Newfoundland 91, 105
Niagara River 91
Niger 192
Niger Delta Development Board (Nigeria) 194
Nigeria 3, 189, 203, 205, 207, 208, 209, 210, 212, 213, 214, 215
Nogoya City 155
non-structural adjustments 26, 28, 102, 149, 164, 205, 209
nonpoint pollution 1, 27, 101, 120, 208
North Island 50, 51, 80
North Sea Oil 127
Northland 72
Northwest Power Planning Council (United States) 36
Northwest Territories 92
Notec River 178
Nova Scotia 105

Oder River 172, 175
Office of Water Resources and Technology (United States) 29
Ogun–Oshun River Basin Development Authority (Nigeria) 199, 200
Ohio Conservation Districts (United States) 2
oil 98, 105, 127, 193
Okanagan River 101
Okutadami special area (Japan) 153
Ontario 93, 102, 103, 105, 210, 213
open space 39, 70
Oregon 36
organizational culture 14–16, 216–217
Osaka 152, 156
Ottawa River 102
Outdoor Recreation Resources Review Commission (United States) 33
over-grazed 50
Oyo State 199, 200
Oyo State Water Corporation (Nigeria) 199

Pacific Southwest Basin (United States) 34
paddy fields 149, 152, 161, 163
parks 97
parochialism 14, 75, 123, 140
participant attitudes 14–16, 216–217
paternalism 63
penalties 66
Perspective Plan for Water Management in Poland 175, 176, 179
Philippines 3
Pick-Sloan Bill (United States) 26
pilot water management systems 178
Poland 172, 203, 206, 207, 208, 210, 212, 213, 214, 215
Polish Academy of Sciences 175, 177, 185
politically expedient 53
pollution 10, 11, 34, 58, 67, 72, 93, 101, 120,

129, 131, 133, 149, 163, 164, 165, 176,
177, 181, 182, 184, 185, 195, 198, 200,
205, 210, 212
Poznan 176
pre-planning 102
preventive 99
pricing 29, 103, 143, 215
Prince Edward Island 91, 105
privatization 78, 121, 132, 133, 134, 135,
141, 142, 205, 206, 208, 210
proactive 211, 212
processes and mechanisms 13–14, 214–215
professional biases 83, 204
professional rivalry 54, 55, 216
property owners 63
property taxes 94
prosecution 66, 132
protectionism 48
provincial-municipal partnership 94, 96
public hearings 68
public ownership 31
public participation 14, 110, 112, 215

Qu'Appelle River 101
Quebec 105

rabbits 50
Rafferty/Alameda project (Canada) 107
Rakaia River 68
rates 53, 54, 55, 63, 68, 71, 83, 127
rational-comprehensive 136
reactive 55, 99
Reagan administration 28, 29, 40, 211
recession 127, 140, 178
recreation 1, 11, 31, 33, 59, 60, 62, 67, 72,
92, 94, 120, 128, 133, 184, 195, 211, 212
regional development 29
regional planning 14, 71, 72
Regional Water Authorities (England) 123,
131, 132
regional water boards 49, 60, 64
reliability 23
Remedial Action Plans (Canada) 111
Republicans (United States) 24, 26, 27
resettlement 164, 195, 198, 205
resource users 15
responsibility 8
Rhine River 150
rice 149, 161
Rio Grande Basin 31
River Basin Commissions (United States)
34
River Basin Development Authorities
(Nigeria) 194, 195, 199, 200, 205, 210,
212, 213
river control 51
Rivers (Prevention of Pollution) Acts, 1951
and 1961 (England and Wales) 121,
128, 129

Roosevelt, Franklin D. 25
Roosevelt, Theodore 24, 25
Round Tables (Canada) 111, 112, 113

sabotage 15
Sahel 189
salinity 4
salmon 37, 135, 184
Saskatchewan 89, 93, 105, 107
Saskatchewan Water Corporation (Canada)
99, 107
satisficing 137
scenery 59, 62, 72, 211
scenic rivers 64, 68
Science Council of Canada 88
scoping 108
Scotland 120, 126
Sea of Japan 149
sea pollution 158
seashore 60
Second National Development Plan, 1970–
1974 (Nigeria) 193, 194
secrecy 129, 131, 137, 139
Severn-Trent Regional Water Authority
(England and Wales) 129, 135
sewage 11, 58, 70, 82, 120, 124, 125, 126,
128, 131, 132, 134, 135, 140, 156, 158,
178, 179, 184, 185, 186, 212
Shebenacadie-Stewiacke River 101
Sheep 50
Shikoku 149
Silesia 176, 178
silting 72
slum housing 123
social impact 26
social well-being 27, 29
Soil Conservation and Rivers Control Act,
1941 (New Zealand) 49, 73, 75, 209
Soil Conservation and Rivers Control
Council (New Zealand) 51
soil conservation districts 2
Soil Conservation Service (United States 31
soil moisture 23
Sokoto-Rima Basin (Niveria) 198
Souris Basin Development Authority
(Canada) 107
Souris River 101, 107
South Island 50, 80
South West Water Authority (England) 120
Special Area Multiple Purpose Development
Projects (Japan) 153
Special Multiple Purpose Dam Law, 1957
(Japan) 154
St. John River 101, 102
St. Lawrence River 102
Stalemate 16
standards 27, 64, 128, 132, 134, 136, 139,
157, 158, 166, 184, 185, 195
State Owned Enterprises Act, 1986 (New
Zealand) 48

structural adjustments 31, 38, 40, 96, 102, 109, 149, 152, 153, 164, 205, 209, 212
structural determinism 138
structures 11–13, 213–214
Sudan 189
super agencies 214
supply management 31
Susquehanna River Basin Commission (United States) 35, 213
sustainable development 1, 111, 170, 204

Tasman Sea 50
taxing powers 30
Tennessee-Tombigbee navigation project (United States) 28
Tennessee Valley Authority 2, 25, 35–36, 153, 158, 203, 213
territorial authorities 69–75
Thames River 135, 138
Thames Regional Water Authority (England) 135
Thatcher government 119, 121, 127, 137, 138, 140, 142, 205, 211
Third National Development Plan, 1975–1980 (Nigeria) 194
Tiga Dam 198
Tokyo 152, 155, 156, 157
Tone River 150
Tone special area (Japan) 153
top down 14, 17, 215
Toronto 89
tourism 11, 98, 120
Town and Country Planning Act, 1953 (New Zealand) 53, 70
Town and Country Planning Act, 1977 (New Zealand) 50, 72, 76, 209
toxic 91, 103, 163, 208
transportation 1, 203
tussock grass 50
Trent River Authorities (England) 123
Typhoon Ione 160
Typhoon Katherine 160
Typhoon no. 15 160
typhoons 149, 160, 163, 208

uncertainty 35, 113, 161
Unified National Program for Floodplain Management (United States) 39
united councils 71
United States 22, 93, 107, 108, 110, 157, 158, 203, 205, 206, 207, 209, 211, 212, 213, 214, 215, 216
Upper Waitemata Harbour 76
urbanization 164, 184, 185, 188, 205
user pays 83

Vancouver 89
vision 15
Vistula River 172, 175, 176

Waikato United Council (New Zealand) 72
Waihou Valley Scheme (New Zealand) 56
Wales 2, 148, 213
Warsaw 176
Washington 36
waste treatment 1, 211
wastes 64, 65, 67, 120, 131, 157, 158, 178, 184
Water Act, 1945 (England) 121
Water Act, 1973 (England) 127
Water Act, 1983 (England) 127, 210
Water and Soil Conservation Act, 1967 (New Zealand) 49, 59–60, 75, 209, 211
Water and Soil Management Plans (New Zealand) 68–69
Water Authorities Association (England) 135
water balance 37, 172
water classification 64–65
water exports 110, 111
Water Law, 1962 (Poland) 177
Water Law, 1979 (Poland) 177
Water Pollution Act, 1953 (New Zealand) 58
Water Pollution Council (New Zealand) 58, 65
Water Pollution Prevention Law, 1970 (Japan) 157
water quality 1, 26, 27, 40, 46, 53, 60, 64, 65, 67, 68, 72, 75, 91, 92, 97, 99, 103, 108, 111, 120, 125, 128–133, 134, 137, 148, 156, 158, 163, 166, 168, 169, 175, 178, 181, 205, 208, 210, 211, 212, 213, 215
Water Resources Act, 1963 (England and Wales) 121, 123
Water Resources Board (England and Wales) 123
Water Resources Council (New Zealand) 64
Water Resources Council (United States) 27
Water Resources Development Corporation Law (Japan) 154
Water Resources Development Law (Japan) 154
Water Resources Planning Act, 1969 (United States) 28, 209
Water Resources Research Act (United States) 29
water re-use 123
water rights 66, 67, 212, 215
water supply 1, 10, 70, 82, 92, 97, 99, 105, 120, 122, 126, 128, 133, 182, 185, 186, 191, 195, 198, 199, 205, 211, 212
Waterworks Law, 1977 (Japan) 156
Wellington 58, 63, 65, 72
wetlands 1, 11, 80, 91, 97, 206
white coal 176
wild and scenic rivers 27, 68

Wild and Scenic Rivers Act, 1968 (United States) 27
wilderness 62
wildlife 4, 11, 31, 32, 33, 36, 59, 60, 62, 67, 72, 76, 80, 108, 133, 206, 211
Winnipeg 89
World Commission on Environment and Development (Brundtland Commission) 1, 111, 170, 204
World Conservation Strategy 110
Wroclaw 176

Yahagi River 149, 164–169, 216
Yahagi River Basin Development Research Association (Japan) 168
Yahagi River Basin Water Quality Protection Association (Japan) 164–169, 205, 215
Yorkshire Water Authority (England) 136
Yosui 152
Yukon 89, 92, 102

zoning 40